**Microfabricated Power
Generation Devices**

*Edited by
Alexander Mitsos and
Paul I. Barton*

D1702292

Further Reading

Kolb, G.

Fuel Processing

for Fuel Cells

2008
Hardcover
ISBN: 978-3-527-31581-9

Ozawa, K. (ed.)

Lithium Ion Rechargable Batteries

Materials, Technology, and Applications
2009
Hardcover
ISBN: 978-3-527-31983-1

Hessel, V., Renken, A., Schouten, J. C., Yoshida, J.-I. (eds.)

Micro Process Engineering

A Comprehensive Handbook
2009
Hardcover
ISBN: 978-3-527-31550-5

Geschke, O., Klank, H., Telleman, P. (eds.)

Microsystem Engineering of Lab-on-a-Chip Devices

Second, Revised and Enlarged Edition
2008
Hardcover
ISBN: 978-3-527-31942-8

Wirth, T. (ed.)

Microreactors in Organic Synthesis and Catalysis
2008
Hardcover
ISBN: 978-3-527-31869-8

Microfabricated Power Generation Devices

Design and Technology

Edited by
Alexander Mitsos and Paul I. Barton

WILEY-VCH Verlag GmbH & Co. KGaA

The Editors

Dr. Alexander Mitsos
RWTH Aachen AICES
Pauwelstr. 12
52074 Aachen
Germany

Prof. Paul I. Barton
Mass. Inst. of Technology
Dept. of Chemical Engineering
66-464, 77 Massachusetts Ave.
Cambridge, MA 02139
USA

Library of Congress Card No.:   applied for

British Library Cataloguing-in-Publication Data
A catalogue record for this book is available from the British Library.

Bibliographic information published by the Deutsche Nationalbibliothek
The Deutsche Nationalbibliothek lists this publication in the Deutsche Nationalbibliografie; detailed bibliographic data are available on the Internet at <http://dnb.d-nb.de>.

© 2009 WILEY-VCH Verlag GmbH & Co. KGaA, Weinheim

Typesetting   SNP Best-set Typesetter Ltd., Hong Kong
Printing   Betz-druck GmbH, Darmstadt
Binding   Litges & Dopf GmbH, Heppenheim
Cover Design   Adam-Design, Weinheim

Printed in the Federal Republic of Germany
Printed on acid-free paper

ISBN   978-3-527-32081-3

Contents

Microfabricated Power Generation Devices. Edited by Alexander Mitsos and Paul I. Barton
Copyright © 2009 WILEY-VCH Verlag GmbH & Co. KGaA, Weinheim
ISBN: 978-3-527-32081-3

Preface

In the past decades there has been a dramatic increase in the power demand for portable electronics, and batteries are in many respects limiting. As a consequence, a variety of new power generation devices are under research and development, such as microfabricated fuel cell systems or microturbines. There are numerous publications on micropower devices, but they only cover aspects of this complicated system. This book is the first to give a broad coverage on portable power generation devices, encompassing technological aspects and system design using computational methods. The chapters are written by internationally recognized experts. Each chapter is self-contained, but at the same time there is a continuity between chapters, ensuring a coherent book as opposed to a collection of articles. We believe that the book is at the same time a great introduction for beginners and a very valuable tool for expert researchers in academic and for-profit institutions. While not a text-book, our own experience shows that it is also suitable as a complement to micropower design courses.

We are grateful to the contributors of the individual chapters; we appreciate the time they devoted in writing articles that describe the state-of-the-art in a tutorial fashion. In many ways the basis of this book was a Multidisciplinary University Research Initiative (MURI) at MIT. Many of our colleagues in the MURI program contributed as authors of individual chapters. We would also like to thank our colleagues who did not have the chance of contributing, in particular: Klavs F. Jensen for his leadership of the MURI; Leonel R. Arana and Steve Weiss for numerous discussions; Michael M. Hencke and Ignasi Palou-Rivera for their contribution to the system-level analysis; Ruth Misener for her numerical experiments on DAEs inside DAEs; and Mehmet Yunt for his work on variable power demand. Finally we thank the publisher for the invitation to put together this book.

This work was supported by the DoD Multidisciplinary University Research Initiative (MURI) program administered by the Army Research Office under Grant DAAD19-01-1-0566. Financial support from the Deutsche Forschungsgemeinschaft through grant GSC 111 is gratefully acknowledged.

June, 2008   *Alexander Mitsos and Paul I. Barton*

List of Contributors

Paul I. Barton
Massachusetts Institute of
Technology
Department of Chemical
Engineering
66-464
77 Massachusetts Avenue
Cambridge
MA 02139
USA

Jürgen J. Brandner
Forschungszentrum Karlsruhe
Institute for Micro Process
Engineering
P.O. Box 3640
76021 Karlsruhe
Germany

Benoît Chachuat
McMaster University
Department of Chemical
Engineering
1280 Main Street West
Hamilton
Ontario L8S 4L7
Canada

Kishori Deshpande
Dow Chemical Company
Engineering and Process Sciences
B-1603
Room 1223
Freeport
TX 77541
USA

Alan Epstein
Massachusetts Institute of Technology
Department of Aeronautics and
Astronautics
Room 31-269
77 Massachusetts Avenue
Cambridge
MA 02139
USA

Edward P. Gatzke
University of South Carolina
Department of Chemical Engineering
301 Main St. Columbia
SC 29208
USA

Microfabricated Power Generation Devices. Edited by Alexander Mitsos and Paul I. Barton
Copyright © 2009 WILEY-VCH Verlag GmbH & Co. KGaA, Weinheim
ISBN: 978-3-527-32081-3

Joshua L. Hertz
University of Delaware
Department of Mechanical
Engineering
126 Spencer Laboratory
Newark
DE 19716
USA

Volker Hessel
Institut für Mikrotechnik Mainz
GmbH
Chemical Micro & Mill
Processing Technologies
Carl-Zeiss-Str. 18–20
55129 Mainz
Germany

Eindhoven University of
Technology
Department of Chemical
Engineering and Chemistry
P.O. Box 513
5600 MB Eindhoven
The Netherlands

Stuart Jacobson
Massachusetts Institute of
Technology
Department of Aeronautics and
Astronautics
Room 31-269
77 Massachusetts Avenue
Cambridge
MA 02139
USA

Gunther Kolb
Institut für Mikrotechnik Mainz
GmbH
Department of Energy
Technology and Catalysis
Carl-Zeiss-Str. 18–20
55129 Mainz
Germany

Mayuresh V. Kothare
Lehigh University
Department of Chemical Engineering
D322 Iacocca Hall
111 Research Drive
Bethlehem
PA 18015
USA

Hanqing Li
Massachusetts Institute of Technology
Department of Aeronautics and
Astronautics
Room 31-269
77 Massachusetts Avenue
Cambridge
MA 02139
USA

Alexander Mitsos
RWTH Aachen
Aachen Institute for Advanced Study in
Computational Engineering Science
Pauwelsstr. 12
52074 Aachen
Germany

Alexander Mitsos
RES Group Inc/Numerica Technology
LLC
4 Cambridge Center
Cambridge
MA 02142
USA

Ole Nielsen
Bose Corporation
The Mountain
Framingham
MA 01701
USA

S. Mark Spearing
School of Engineering Sciences
University of Southampton
Highfield
Southampton
SO17 1BJ
U.K.

Andrew T. Stamps
Air Products and Chemicals
7201 Hamilton Boulevard
Allentown
PA 18195-1501
USA

Harry L. Tuller
Massachusetts Institute of
Technology
Department of Materials Science
and Engineering
Room 13-3126
77 Massachusetts Avenue
Cambridge
MA 02139
USA

Dionisios G. Vlachos
University of Delaware
Department of Chemical Engineering
and Center for Catalytic Science and
Technology
Newark
DE 19716-3110
USA

Brian L. Wardle
Massachusetts Institute of Technology
Department of Aeronautics and
Astronautics
Building 33-314
77 Massachusetts Avenue
Cambridge
MA 02139-4307
USA

Benjamin A. Wilhite
University of Connecticut
Department of Chemical
Materials and Biomolecular
Engineering and Connecticut Global
Fuel Cell Center
191 Auditorium Road
Unit 3222
Storrs
CT 06269
USA

1
Introduction

Alexander Mitsos and Paul I. Barton

The widespread use of portable electric and electronic devices increases the need for efficient autonomous man-portable power supplies, of the order of 0.1 W to about 50 W. The predominant technology for portable power generation is the battery. However, the energy density of batteries is of the order of only a few hundred $Wh\,l^{-1}$ and $Wh\,kg^{-1}$. Battery performance has significantly improved over the last decades, but it is believed that the upper limit on performance is being approached, because the list of potential materials is being depleted. Additionally, batteries have high cost and life cycle environmental impact.

This book focuses on alternatives to batteries based on microchemical systems, that is, miniaturized devices that use chemical fuels as the primary source of energy. Energy conversion technologies considered include electrochemical reactions in fuel cells, combustion in connection with heat-engines and combustion combined with thermo-photovoltaic elements. The promise of these systems is that significant increments in energy density can be made compared to state-of-the-art batteries. Suppose, for the sake of argument, that it would be practical to operate a fuel cell reversibly at ambient conditions. Then, common chemicals would outperform state-of-the-art batteries by orders of magnitude, as shown in Figure 1.1 [1]. As a consequence, relatively inefficient microchemical systems can significantly outperform batteries. Miniaturization has the advantage of smaller systems as well as potential process intensification. Such considerations have sparked great military [2] and civilian interest in developing these alternatives to batteries.

1.1
Alternatives to Microchemical Systems

The focus of this book is on microfabricated devices that transform chemical energy to power, either directly (fuel cells) or indirectly (micro-engine, photovoltaics). This section briefly describes some other potential alternatives.

Microfabricated Power Generation Devices. Edited by Alexander Mitsos and Paul I. Barton
Copyright © 2009 WILEY-VCH Verlag GmbH & Co. KGaA, Weinheim
ISBN: 978-3-527-32081-3

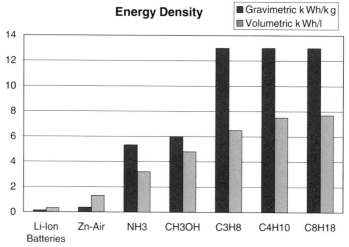

Figure 1.1 Comparison of state-of-the-art batteries with theoretical energy density of fuels in a reversible fuel cell at ambient temperature, in which all the Gibbs free energy of oxidation is converted to power.

1.1.1
Using Manpower

Arguably the most readily available and simplest to use energy form is the mechanical work that the human body can generate. Two main alternatives exist, namely the transformation of active work (e.g. turning a foot pedal), and the capture of passive work (e.g. energy dissipated while running). The human body transforms chemical energy at a rate of the order of a few hundred Watts and therefore capturing a small fraction of this energy could power some man-portable electronic devices. It is interesting to note that some energy-dense food has energy densities comparable to the chemicals considered in the following chapters. Moreover, the body's energy conversion efficiency is approximately 25% [3, 4], which makes it competitive with some of the technologies considered in this book. However, using the body's energy is not as comfortable as using a micro-chemical system and is therefore expected to find use only in niche applications. Various technologies have been proposed in the past to capture and transform this energy:

1. Dynamos which convert mechanical rotation into a direct electric current. Dynamos use the electromagnetic principle: when an electrical conductor moves in a magnetic field, a potential difference is generated.

2. Mechanical systems, for example, by storing the kinetic energy in a spring. The stored mechanical energy is then used for electricity production.

3. Piezoelectric systems which rely on the ability of certain materials to generate an electric potential in response to applied mechanical stress.

Some researchers also consider the possibility of using the heat released by the body (thermoelectric conversion). Since body heat is available at very low temperatures, the efficiency would be very low, even if the Carnot limit was reached. For instance, at an ambient temperature of $T = 298$ K the maximal efficiency would be $\frac{310-298}{310} = 4\%$.

Well-known applications include low-tech devices such as bicycle lights and rechargeable flash lights. In the "One laptop per child" initiative [5] a foot pedal is considered as an alternative for the powering of laptops. People have also envisioned high-tech devices such as shoe generators; while running, significant amounts of energy are dissipated in the human tissue as well as in the middle sole of modern running shoes; therefore, a long term goal of ingenious inventors has been capturing part of this dissipated energy and several patents have been granted [6]. A similar idea is the suspended-load backpack [3, 4] and the knee brace [7].

1.1.2
Harvesting Environmental Energy

An appealing alternative for some applications is harvesting energy from the environment. The most common option is using photovoltaics, which are very practical for powering low power consuming devices such as pocket calculators. However, current technologies do not seem promising for power-intense man-portable devices such as laptops, because the required surface area would be very big (of the order of few m^2).

Recently, the concept of wireless power transfer has been revisited. For instance, a research group at MIT has demonstrated the transfer of 60 W over 2 m with 40% efficiency [8]. The promise of wireless power transfer is to eliminate the need for cables, transformers and so on, and, therefore, it is not truly an alternative to man-portable power generation, but rather to the recharging of batteries. A similar concept is collecting vibrational energy, for example [9].

1.2
Book Outline

The remainder of the book is organized around two main themes. In Part One the various technologies are discussed by experts in the respective fields. First, an introduction to microfabrication is given, focusing on functionalization such as catalyst coating. Then, the various potential components of microchemical systems for portable power generation are discussed, namely fuel processing, fuel cells, micro heat engines and thermo-photovoltaics. Part One concludes with a discussion of system integration from the technology point of view, covering issues such as thermal and material management.

Part Two discusses the state-of-the-art in system design using computational methods. First the selection of alternatives is discussed based on simple algebraic models. Then material and structural considerations are covered. In Chapter 10 methodologies for reactor and reaction engineering are presented based on multi-scale modeling. In Chapter 11 the optimal design and operation of fixed alternatives are discussed based on models of intermediate fidelity. Chapter 12 discusses challenges in hybrid electrochemical systems, for example, the combination of a fuel cell with a battery. The book concludes with a discussion of process control issues.

References

1 Mitsos, A., Palou-Rivera, I. and Barton, P.I. (2004) Alternatives for micropower generation processes. *Industrial and Engineering Chemistry Research*, **43** (1), 74–84.

2 National Research Council Committee of Soldier Power/Energy Systems. (2004) *Meeting the Energy Needs of Future Warriors*, National Academy Press, Washington, D.C.

3 Kuo, A.D. (2005) BIOPHYSICS: harvesting energy by improving the economy of human walking. *Science*, **309** (5741), 1686–7.

4 Rome, L.C., Flynn, L., Goldman, E.M. and Yoo, T.D. (2005) Generating electricity while walking with loads. *Science*, **309** (5741), 1725–8.

5 One laptop per child http://laptop.org/ (accessed May 29, 2008).

6 Paradiso, J.A. (2006) Systems for humanpowered mobile computing, Proc. of the IEEE Design Automation Conference (DAC2006), San Francisco CA, July 24–6, pp. 645–50.

7 Donelan, J.M., Li, Q., Naing, V., Hoffer, J.A., Weber, D.J. and Kuo, A.D. (2008) Biomechanical energy harvesting: generating electricity during walking with minimal user effort. *Science*, **319** (5864), 807–10.

8 Kurs, A., Karalis, A., Moffatt, R., Joannopoulos, J.D., Fisher, P. and Soljacic, M. (2007) Wireless power transfer via strongly coupled magnetic resonances. *Science*, **317** (5834), 83–6.

9 Roundy, S., Wright, P.K. and Rabaey, J. (2003) A study of low level vibrations as a power source for wireless sensor nodes. *Computer Communications*, **26** (11), 1131–44.

Part One   Technologies

2
Microfabrication for Energy Generating Devices and Fuel Processors

Volker Hessel, Gunther Kolb and Jürgen J. Brandner

2.1
Specific Issues for Manufacturing of Energy-Related Microdevices

The microfabrication of energy generating and converting microdevices is different from their chemistry oriented micro counterparts. Chemical microreactors are frequently fabricated as few or even single devices. In contrast, the development of microdevices and microreactors, as well as fuel processors for future distributed energy generation, heads for mass production. For chemical microreactors, the figure of merit is their level of process intensification, since this relates massively to the capital (CAPEX) and operational (OPEX) costs of the corresponding micro-chemical plants. For fuel processors and related energy generating/converting devices energy efficiency and reactor fabrication costs are the most crucial issues. For chemical microreactors even higher fabrication (CAPEX) costs may be tolerated if this is outperformed by the process intensification benefits which reduce the OPEX costs. The same is hardly true for the microfabrication under consideration in this book.

2.2
Micro Power Device Type and Choice of Microfabrication Techniques – Role of Materials, Sealing, and Dimensions

The manufacturing technologies can be grouped into two different classes for all the materials considered here, that is erosive and generative manufacturing. Technologies like embossing or molding belong to the class of generative techniques, while laser micromachining, for example, is an abrasive technique; more forming the material than really manufacturing it.

Micro power devices, generally microreactors, comprise a (sometimes heterogeneously composed) family of new compact, energy-efficient reactors with distinct sub-classes like monoliths, foams, plate heat-exchangers/reactors, microstructured reactors, or chip-based systems. Each of these sub-classes typically has

Microfabricated Power Generation Devices. Edited by Alexander Mitsos and Paul I. Barton
Copyright © 2009 WILEY-VCH Verlag GmbH & Co. KGaA, Weinheim
ISBN: 978-3-527-32081-3

its own microfabrication techniques, both by tradition and by real need, and accordingly the following sections are grouped in that way. Some of these types of reactors can be made out of different materials with distinct reactor construction and application, which again leads to different fabrication techniques, as given for ceramic or metallic monoliths.

Besides the pure fabrication of the microstructures, the type and way of sealing the microstructured units to a system varies. Initial test systems are typically reversibly sealed for ease of opening and inspection. When going to real-case designs the more compact irreversible sealing is preferred. Different techniques are available for both approaches.

Steel and metals are today's preferred materials for the (larger) microstructured reactors, while polymers and silicon have widely established microfabrication technologies, stemming from the microelectronics area, and are used for making the (smaller) chip systems. Ceramics and glass are niche materials for high-temperature applications and to provide environments with zero catalytic blind activity; in addition, glass materials allow visual inspection.

For steel, polymers, and silicon mass-manufacturing production lines are available in principle, but certainly need adaptation to the specific microdevice to be manufactured. In particular, automated device manufacture needs to be complemented by automated catalyst coating. Concerning the latter, more development is needed and only the first steps have been undertaken. A particular issue is to integrate functions in the microdevice such as the separation of gas components by means of membranes. The latter again demands smart microfabrication.

2.3
Fabrication Techniques for Minichannels in Monoliths or Mini Fluidic Ducts in Foams

Monoliths are porous blocks with a multitude of parallel minichannels and can be considered as the slightly larger counterparts of microreactors of the same architecture. However, complex fluidic arrangements such as counterflow heat-exchanger-reactor designs are not as commonly found for monoliths as for micro-reactors, due to the limitations of their fabrication techniques.

2.3.1
Fabrication of Ceramic Monoliths

Ceramic monoliths usually consist of cordierite as carrier and construction material [7]. In some cases, alumina is used. Cordierite is mixed with water and binder material such as methylcellulose to give a paste. Heat resistant inorganic fibers may be added to improve the mechanical properties of the paste for further processing [8]. The paste is then shaped by extrusion to a monolith green body and dried and calcined at 1300–1400 °C for 3 to 4 h [7]. Instead of the often not readily available cordierite, formulations of source materials may be mixed to match the composition of cordierite ($5SiO_2 : 2Al_2O_3 : 2MgO$).

After coating with catalyst, the ceramic monolith, surrounded and fixed by a ductile mat material, is encased in a metallic reactor shell for reasons of mechanical stability, prevention of gas by-passing, and interconnection to the fluidic environment [9]. Interam, produced by 3M, is a ceramic mat used in automotive exhaust systems and can be used in the medium temperature range [9]. For applications operating at a temperature above 800 °C, which is common for monolithic reformer reactors, high temperature ceramic fiber materials, such as CC-Max[@] from Unifrax, have to be used [9].

2.3.2
Fabrication of Metallic Monoliths

Metallic monoliths consist of rolled-up flat and corrugated metal foils with a thickness as low as 20 μm and composed of alloys of iron, with about 15–20 wt.% chromium and 5 wt.% aluminum (Fecralloy). Alternating arrangement of both foils can be achieved (see Figure 2.1).

Fecralloy alloys form a 0.5 μm thin alumina outer layer when thermally treated above 850 °C which acts as an adhesion layer for catalyst coating and a protective layer for corrosion resistance [7]. Increasing the aluminum content above 6 wt.% increases the resistance of the material against corrosion, but also increase its brittleness. For strengthening of axial mechanical stability, forcing bins, welding and brazing techniques have been applied [7]. Sandwich foils of Fecralloy and aluminum can be welded during operation [7]. Different metallic monolith structures have been constructed in this way by the EMITEC Company. Alternatively, stacks of corrugated and flat foils are wound around two ("S"-type in Figure 2.2) and three ("SM"-type in Figure 2.2) mandrels. Torsional deformation then stabilizes the monolith structure.

Aluminum is an alternative construction material for ceramic monoliths but has a lower operating maximal temperature of about 450 °C.

Figure 2.1 View into metallic monoliths produced by EMITEC (photo: courtesy of Emitec).

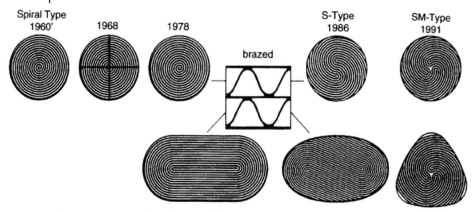

Figure 2.2 History of metallic monolith designs as developed by EMITEC [7].

2.4
Fabrication of Ceramic and Metallic Foams

Ceramic foams are made, for example, from alumina, alumina silicates, zirconia, stabilized zirconia and titania. Organic foams such as polyurethane or polyolefins constitute the negative precursor structure of the ceramic foams. These polymer foams are filled with a 20 wt.% aqueous slurry of the ceramic comprising particles of size 0.1–10 µm, with the aid of wetting agents, dispersion stabilizers and viscosity modifiers [10]. Drying and calcination of the filled foam at 1000 °C leads to combustion of the polymer and sintering of the ceramic.

Metallic foams have similar properties to ceramic foams but superior mechanical stability and improved heat conductivity.

2.4.1
Fabrication of Plate Heat-Exchangers/Reactors

Plate heat-exchangers comprise a stacked arrangement with a number of parallel minichannels and have high surface-to-volume ratios in the range $200 \, \text{m}^2 \, \text{m}^{-3}$ [11].

2.4.2
Steel and Metal Material Choice

The preferred construction material is stainless steel. For high-temperature applications, as required for hydrocarbon reforming, nickel-based alloys or Fecralloys are an alternative. A disadvantage of the latter material is its brittleness. Ceramics are another class of high-temperature materials.

For low-temperature processes such as methanol reforming and carbon monoxide purification, copper and especially aluminum are suited. Their higher, by orders of magnitude, heat conductivity compared to stainless steel provide other options for the thermal management, for example to achieve isothermal conditions, as needed for evaporators or reactors operating within narrow temperature windows. However, aluminum is a less preferred material when considering corrosion issues.

The choice of material for plate heat-exchanger/reactors depends also on the desired dynamic properties of the microsystem. One important figure is the energy demand for fuel processor start-up, which results from the product of specific heat capacity and density of the construction material. For a given geometry and volume of the device, aluminum is favored over copper and stainless steel.

Polymers are promising materials for low-temperature plate heat-exchangers and especially for withdrawal of the last portion of energy out of the fuel processor off-gases before releasing them to the environment.

2.4.3
Steel and Metal Plate Sealing Techniques

Plate heat-exchangers are made by punching, embossing, etching or rolling. Details can be found in Section 2.5 where these methods are given in the example of making microstructured reactors.

An important issue in the highly parallel, stacked multi-plate device approach is to ensure proper sealing. Sealing by gaskets is the method of choice for interconnection and stacking of the plates and also facilitates catalyst coating and cleaning due to easy disassembling. Normally used gasket materials such as nitrile, neoprene and viton limit the operating temperature to 200 °C, as given in Ref. [11]. Above 200 °C, metallic gaskets are the choice. A disadvantage of the concept of sealing by gaskets is the need to insert screws and thus to have a bulky housing. Especially for smaller devices, the thermal mass is increased considerably, which, for example, increases start-up time.

Irreversible sealing techniques typically make use of elevated temperatures for which compatibility with the plate material and its catalyst coatings has to be tested. If not compatible, the catalyst or the catalyst coating has to be inserted into the device after the sealing procedure.

2.4.3.1 Laser Welding
Laser welding is largely applied for sealing of plate heat-exchanger/reactors. Cost issues may arise for devices of larger scale since the length of the weld increases correspondingly. The spatially limited energy input protects incorporated catalyst from damage. Figure 2.3 shows a 10 kW counter-flow heat-exchanger which was sealed by laser welding.

2.4.3.2 Electron Beam Welding
Electron beam welding, established in the automotive industry already, is an alternative with similar locally limited energy input. In contrast, diffusion bonding requires high temperatures with no spatial restriction and also high vacuum. The

Figure 2.3 10 kW counter-flow heat-exchanger sealed by laser-welding (source: IMM).

material is compressed and heated to temperatures close to the melting point, which generates a quasi-boundary-free single workpiece.

2.4.3.3 Diffusion Bonding

Diffusion bonding of nickel-based alloys (Hastelloy) was used for manufacture of a mini-channel (1×1 mm^2) plate and fin heat-exchanger (40 mm $\times$ 40 mm $\times$ 3 mm in size) [12]. After removal of the oxidation layer, compression was carried out at an operating pressure of 6 mPa. For NiCroFer and other materials, an even lower pressure of 1 μPa is satisfactory. Another procedure relies on 1150 °C bonding temperature and 440 bar contact pressure for 30 min [12]. Only 3% deformation was needed for bonding. Leak tightness was given at ambient temperature up to 630 bar. Diffusion bonding was also performed with Fecralloy plates at 900 °C and a pressure of 200 bar [13].

A proprietary diffusion bonding process for fabrication of large meso-scaled plate-type heat-exchangers with exchange capabilities in the MW range was developed by the Heatric Company (www.heatric.com). These meter-long heat-exchangers are used for off-shore applications. The Karlsruhe Research Center practises diffusion bonding frequently and since almost two decades for their cross-flow microstructured platelet heat-exchangers of several cm outer dimensions (see Figure 2.4) [14, 15].

2.4.3.4 Brazing

Brazing techniques also find use for sealing compact plate heat-exchangers. The brazing lots often contain heavy metals such as cadmium and tin, poisonous to catalysts [9]. Thus, the brazing step should be followed by a thorough cleaning step before applying catalyst coating [9]. However, if brazing is done with catalyst-coated plates, the same temperature considerations as given above with regard to the catalyst coatings hold for the melting temperature of brazing lots.

Sintering has been used for sealing a compact methanol fuel processor [16].

(a)

(b)

Figure 2.4 (a) Diffusion bonded cross-flow heat-exchanger bodies developed by Karlsruhe Research Center; (b) ready-made heat-exchangers (source: FZK).

2.5
Fabrication of Microchannels for Microstructured Reactors

In the following discussion of fabrication techniques for microstructured channel systems future mass production capability will be a focus.

2.5.1
Steel and Metal Plate Micromachining

Metals and metal alloys ("steels") are preferred materials for conventional devices in chemical and energy related process engineering, and have a major share as well in microprocess technology. The materials used range from noble metals like gold, silver, platinum, rhodium or palladium via stainless steel to copper, aluminum or Ni-based alloys [1, 4, 5].

2.5.1.1 Mechanical Precision Machining
Mechanical precision machining comprises milling, drilling, slotting and planing, comparable to the techniques well known in conventional dimension machining. The choice of the tools for machining depends on the stability of the material. For brass and copper, natural diamond microtools are suitable, while for stainless steel and nickel-based alloys hard metal tools are needed. In Figure 2.5, a natural diamond cutter is shown, while Figure 2.6 shows a hard metal drill. For all techniques, details can be found in Refs [1, 4, 17–23].

With copper or brass as base material, mechanical precision machining gives the high surface quality of the micromachined structures, when followed by an electropolishing step (roughness down to 30 nm). Figure 2.7 shows the surface of a single copper microchannel after the electropolishing step.

Micromilling is a useful fabrication technique ready at hand for experimental work and rapid prototyping, but does not satisfy future mass production issues.

Figure 2.5 Natural diamond tool for precision micromachining.

Figure 2.6 Hard metal drill for precision micromachining.

Figure 2.7 Microchannel machined in copper with very high surface quality.

2.5.1.2 Micro Electro Discharge Machining

Micro electro discharge machining (μEDM) is controlled spark micromachining between a conductive electrode and a conductive abraded working piece under a dielectric fluid [24, 25]. In one variant, μEDM die-sinking, an electrode is moved close to a workpiece creating a reverse-shaped ablation. This allows the manufacture of complex shapes as, practically, the shape complexity of the electrode determines that of the microstructure. In μEDM wire cutting, the discharge process is achieved by a rotating wire with a (currently) minimum diameter about 20 μm moving through the workpiece. μEDM techniques are suited for rapid prototyping, but not (directly) for mass production of microstructured devices. However, since mold inserts can be fabricated in this way, μEDM opens indirectly the latter route by enabling injection molding, leading most prominently to polymeric microstructures made in high numbers, but also to metallic or ceramic ones.

2.5.1.3 Wet Chemical Etching

Etching techniques were initially developed for silicon micromaching and are suited for mass production. For many metals etching is also a cheap and well established technology [1–5, 26]. It is competitive for mass production and covers a wide range of channel depths from about 100 μm up to 600 μm, which is the channel size usually applied in microstructured reactors for fuel processing applications.

The fabrication sequence starts with coating a photosensitive polymer mask material onto the steel substrate. The mask is exposed to light via a structure primary mask. After development of the polymer, etching is performed to remove material, typically relying on an iron trichloride solution. Details of the etching processes and etchants can be found in Refs [1, 4, 7, 27].

The aspect ratio is limited to <0.5 at the optimum for wet chemical etching. Due to the isotropic wet chemical etching, the minimum width of a structure is two times the depth, plus the width of the mask openings. Dry etching technologies like reactive ion etching (RIE) are suited to reach aspect ratios >0.5, but are rarely applied to machine metal microstructures.

Wet chemical etching yields semi-elliptic or semi-circular microstructures with quite high surface roughness (in the range of some microns), again due to the isotropic etching. Figure 2.8 shows a stainless steel microchannel structure manufactured by wet chemical etching.

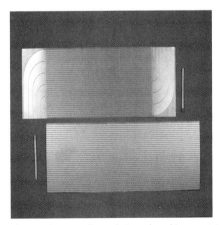

Figure 2.8 Microchannels introduced into a stainless steel plate by wet chemical etching (source: IMM).

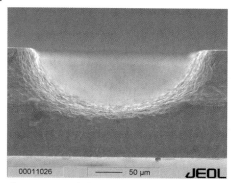

Figure 2.9 Detailed view of a cut etched microchannel in stainless steel with semi-elliptic shape (source: FZK).

The semi-elliptic shape of the etched microstructure is shown in Figure 2.9.

Etching is part of the so-called lamination process using shim- or sheet-technology, being suited for mass production and cost effective [28, 29]. Patterned thin strips are stacked and bonded [30]. The use of spark erosion techniques and punching for manufacturing of single lamination sheets is given in Ref. [31].

2.5.1.4 Selective Laser Melting

A very special method to manufacture microstructures is selective laser melting (SLM) [32–34]. It is one of the rare generative methods with potential for rapid prototyping. A metal powder is distributed on a base platform made of the desired metal material. A focused laser beam is guided along structure lines and melts the metal, forming a welding bead. The position of the platform is then lowered for new distribution of powder and the process repeated in the style of a layer-by-layer build-up of the microstructure. Figure 2.10 shows a schematic of the working

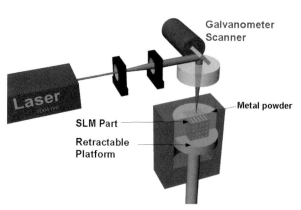

Figure 2.10 Scheme of the SLM process (source: FZK).

(a) (b)

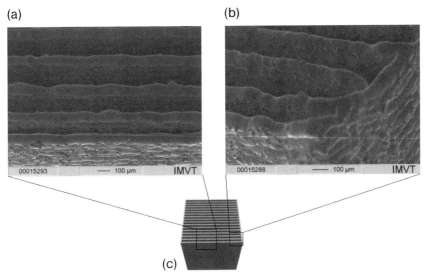

(c)

Figure 2.11 CAD model of a microstructure cube (c), SEM photos of the inner (a) and outer (b) walls of this cube, manufactured by SLM (source: FZK).

principle of this technique, while Figure 2.11 shows a picture of a microstructure stainless steel cube manufactured by SLM.

2.5.1.5 Punching

Punching is an inexpensive but less frequently applied technique for making microchannels, suitable for mass production [35]. Punching usually does not allow one to make grooves or channels on substrates; rather breakthroughs are opened by the punching action due to complete material removal at the exposed sites. Accordingly, unstructured plates need to be inserted in between the punched plates, as kinds of top and bottom plates for each channel array, to achieve a sealed microchannel system. The ability for the production of holes and breakthrough generation allows an easy means for fluid distribution to connect the various plates, besides pure channel manufacture.

For punching, the manufacture of a precise negative model is needed by, for example, precision machining or spark erosion techniques. Structures of different heights enable the formation of geometrically complex microstructures with holes, slits and openings, combined with more conventional structures like channels and voids, within a single working step.

Particular care has to be taken during the punching process to avoid any bends around the recesses created; otherwise a tight sealing will be difficult to manage. Coating of punched microdevices with catalyst has hardly been explored and remains a challenging task.

(a)   (b)

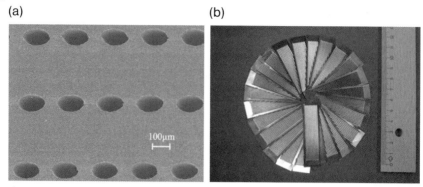

Figure 2.12 (a) Cross-section of a microstructured and embossed heat-exchanger (source: FZK). (b) Several platelets with embossed heat-exchanger microstructures (source: IMM).

2.5.1.6 Embossing

Embossing is another inexpensive technique for manufacturing metal foils, highly suitable for mass production (see Figure 2.12) [36, 37].

Even microstructures down to a few tens of micrometers in size can be achieved by embossing [22]. In Figure 2.13 an embossed microchannel metal foil is shown.

2.5.1.7 Rolling

Corrugated metal foils fabricated by rolling are already applied in the field of automotive exhaust gas systems [38]. The metallic monoliths fabricated this way have channel dimensions in the sub-millimetre range and can be truly termed "microtechnical" devices.

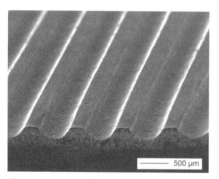

Figure 2.13 Microchannels manufactured in a stainless steel foil by embossing (source: FZK).

2.5.1.8 Laser Ablation

Laser ablation is a frequently applied fabrication technique of proven industrial suitability [39, 40]. However, fabrication of microchannels of several hundred micrometers depth, as is typical for many applications using microstructured reactors, takes too long and thus the method is not cost competitive. For smaller channel dimensions, laser ablation is a viable option, especially for small scale applications.

2.5.1.9 Powder Sintering

Microstructured plates can be formed by sintering of metal powders, as shown for an integrated autothermal methanol reformer being first compressed at a pressure of 1000 bar to shape to a plate-type structure [16]. Sintering of copper, for example, was carried out at 500–700 °C to bond the plates to a stack-like reactor.

Inexpensive low-temperature co-fired ceramic (LTCC DuPont 951 AX) tapes can be used for microreactor fabrication by different routes, via laser machining, milling, chemical treatment, or photolithography [41]. The tapes can be stacked and co-fired to form monoliths. LTCC tapes allow the integration of conducting patterns at their surface, for example for creating resistors and conductors. The smallest tape thickness is 111 µm. The tape composition is oxide particles, glass frit and organic binder. Upon firing, the organic binder combusts and the oxide particles are joined by sintering. The use of LTCC tapes for the construction of a methanol fuel processor is also reported in Ref. [42].

2.5.2
Steel and Metal Plate Interconnection and Assembly

2.5.2.1 Alignment

Correct alignment of the microstructured plates to form stacks or other arrangements is crucial, since only a small distortion while bonding may lead to severe deviations from the ideal microchannel shape. Alignment techniques can be based on simple mechanical methods (e.g. use of alignment pins), edge-catches in a specially designed assembling device or optical methods. Most of these methods stem from silicon micromanufacturing, with precise alignment of multiple mask layers [1, 3]. At the microscale, burr formation from mechanical micromachining or laser machining has to be avoided or minimized, otherwise misalignment of the microstructures will occur.

2.5.2.2 Bonding Techniques

Bonding of metals can be achieved by welding (laser, e-beam, WYG), brazing, diffusion bonding and either low-temperature or high-temperature soldering (P. Pfeifer *et al.*, "*Micromotive*", 2004, unpublished results; W. Pfleging, H. Lambach, unpublished results) [1, 2, 4, 5, 43–59]. Even gluing and clamping have been used. The bonding technique has to be matched to the process parameters of the application of the microdevice, especially the temperatures of the bonding and application processes.

2.5.3
Ceramic Plate Micromachining

Microstructure devices made from ceramic can be applied at up to more than 1000 °C, have no catalytic blind activity, and allow the integration of catalytically active materials. Microfabrication of ceramic microstructures, however, is less developed than for steel and metal micromachining (see below). Experiences obtained with macroscopic devices cannot be used as guidelines directly down to the microscale [6].

Since many ceramic-based routes rely on the use of ceramic powders, the grain size (including its coarsening after sintering) has to be small enough to reproduce precisely all details of the desired microstructure. As a rule of thumb, the grains should be at least one order of magnitude smaller than the smallest dimension of the device. Removing additives during sintering or other processing steps may lead to distortions and cracks, or even to de-binding of microscopic parts of the microdevice

Further, it is not advisable to simply transfer the design of metallic or polymer microdevices to ceramics. Special needs for sealing, assembling and joining as well as for interconnections to metal devices have to be considered.

2.5.3.1 Slurry-Based Molding and Other Techniques
A common route for ceramic microstructures is to prepare a feedstock or a slurry, fluid or plastic molding, injection molding or casting (CIM, HPIM, tape casting), de-molding, de-binding and sintering [6, 50–61]. The shrinkage during the sintering process demands a certain dimensional tolerance. Solid freeform techniques like printing, fused deposition or stereo lithography have also been applied with ceramic slurry. There are certain ceramic materials which can be mechanically machined. In Figure 2.14 ceramic microstructure devices are shown [62].

2.5.3.2 Selective Laser Melting
Selective laser melting was also tried recently for ceramic microstructuring. The first preliminary experiments show feasibility [34].

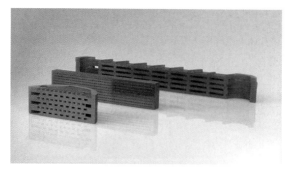

Figure 2.14 Ceramic microstructure devices (Source: ESK, Kempten, Germany, www.esk.com).

2.5.4
Ceramic Plate Interconnection and Assembly

Joining of ceramics to each other or, even more demanding, to metals should be performed by applying materials with similar properties, especially the thermal expansion coefficient. Figure 2.15 shows an example of a ceramic microstructure device connected to metal flanges and fittings.

If joining is done in the green state, firing binds the ceramic material tightly together to form a single body. Another option is soldering with, for example, glass-ceramic sealants. This may limit the working temperature of the microdevice. Reversible assembling and sealing with clamping technologies or gluing as well as conventional seals like polymer o-rings or metal gaskets have also been applied. Details can be found in Refs [1, 6, 60–63]

2.5.5
Polymer Plate Micromachining

Polymer materials have also found use in conventional process engineering and for biological and analytical microdevices, but are much less common for the energy-related microdevices under consideration here [64–68]. Especially with regard to the latter, limitations exist for the maximal operating temperature, although some high-tech polymer materials can withstand temperatures up to 400 °C. The pressure and chemical stability of the polymer microdevices may also be poorer than some of the other materials mentioned above. The integration of catalytic active sites is also less developed. However, polymers are attractive due to the availability of cheap mass production routes, especially by injection molding (see below). The optical transparency is useful for analytical investigations.

Machining and laser ablation are abrasive techniques, while embossing, injection molding, and micro stereo lithography are generative. The most commonly used techniques are injection molding and hot embossing. These two techniques

Figure 2.15 Ceramic microstructure device connected to flange fittings (source: ESK, Kempten, Germany, www.esk.com).

are well established for mass and serial production for both conventional and micro-manufacturing [1, 6, 64–68].

2.5.5.1 Injection Molding

Polymer is the preferred material for injection molding. Microchannels have been made by micro injection molding for numerous applications, for example to form micro heat-exchangers [69].

Embossing is another widely used manufacturing technique for polymer micro-structuring [70–72]. Injection molding and embossing are mass manufacturing techniques.

2.5.5.2 Laser Machining and Micro Stereolithography

Laser machining of polymer microstructures can be done using Excimer or UV-Nd:YAG lasers pulsed at high frequencies [73–75].

Micro stereolithography can be used for the generation of small numbers of prototypes [76–79]. A layer-by-layer generation is achieved through the exposure of a photosensitive monomer plastic spread on a base platform using a focused low power laser. After the generation of one layer, the base platform is lowered to be flooded with the monomer again. In Figure 2.16, a microstructure device made by micro stereolithography is shown. The polymer used here is stable up to a temperature of about 100 °C and is more or less optically transparent.

2.5.6
Polymer Plate Interconnection and Assembly

Gluing and welding have been applied for polymer bonding [6, 80–88]. The gluing method depends heavily on the polymer used. Welding is much more flexible in material choice; ultrasonic, laser or solution welding, for example, can be performed with most of the available polymer materials. In Figure 2.17, a stack of polyvinylidene fluoride foils is shown after a welding step.

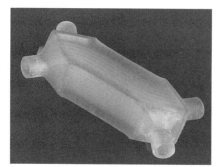

Figure 2.16 Counter-flow design microstructure polymer heat-exchanger made by micro stereolithography (source: FZK).

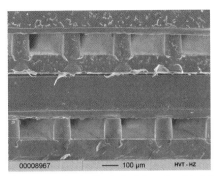

Figure 2.17 Stack of polyvinylidene fluoride foils after a welding step. Welding was not performed correctly, thus the stack shows a gap right in the center of the structures (source: FZK).

2.6
Fabrication of Microchannels for Chip-Like Microreactors

Machining techniques for the production of chip-like systems differ typically from those for microstructured reactors as given above, mainly since the materials differ, but also because the demands on the degree of miniaturization and functionalization/integration are different. These techniques will, most prominently, lead to the future manufacture of small and smallest scale (watt and sub-watt) fuel processors [89–91].

2.6.1
Silicon Micromachining

Silicon micromachining is a largely established technology, stemming from microelectronics, with a number of sequential processing steps. Two major variants are bulk and surface micromachining, typically used for the manufacture of microelectromechanical systems (MEMS) [1–3].

Bulk micromachining defines structures by selectively etching inside a substrate. Usually, silicon wafers are used as substrates. They are exposed by a photolithography masking step to transfer a pattern from a mask to the surface and then anisotropically wet etched by alkaline liquid solvents, such as potassium hydroxide or tetramethylammonium hydroxide (TMAH), forming highly regular V-shaped structures. The surface of these grooves can be atomically smooth. In a fewer number of cases, bulk micromachining has also been performed with dry etches. Dry etching can combine chemical etching with physical etching, or ion bombardment of the material.

Unlike bulk micromachining, surface micromachining is based on the deposition and etching of different structural layers on top of the substrate [92]. As the structures are built on top of the substrate and not inside it, the substrate's properties are not as important as in bulk micromachining. Thus, less expensive substrates such as glass or plastic can be applied.

In surface micromachining layers are grown on top of a silicon substrate. These layers are selectively etched by photolithography using different masks and either a wet or dry etch. Complicated components, such as movable parts (cantilevers), are built using a sacrificial layer which is removed during the fabrication process and gives room to deposit a structural layer.

The use of silicon micromachining to fabricate a chip-like microreactor from a silicon wafer is exemplified in the following (see Figure 2.18) [93]. First, a 10 μm thin photoresist layer was coated onto the silicon chip, to be used as an etch mask for deep reactive ion etching (DRIE). Then, the photoresist was removed and the

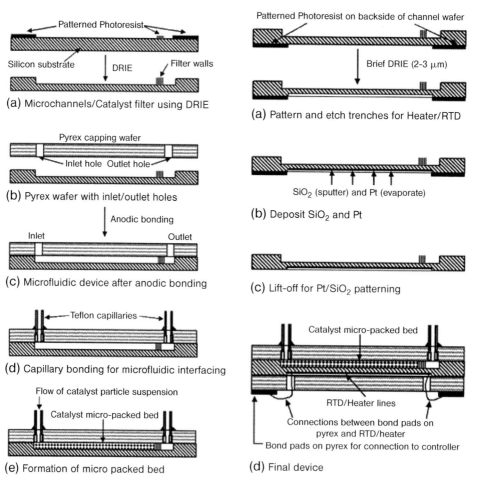

(a) Microchannels/Catalyst filter using DRIE

(b) Pyrex wafer with inlet/outlet holes

(c) Microfluidic device after anodic bonding

(d) Capillary bonding for microfluidic interfacing

(e) Formation of micro packed bed

(a) Pattern and etch trenches for Heater/RTD

(b) Deposit SiO$_2$ and Pt

(c) Lift-off for Pt/SiO$_2$ patterning

(d) Final device

Figure 2.18 Fabrication steps of a MEMS-type methanol steam reformer by deep reactive ion etching [93].

chip was connected to Pyrex glass by anodic bonding to seal the channels. Teflon capillaries were then bonded as inlet/ outlet connectors to the chip. On the reverse side of the chip a photoresist was patterned, which served as an etch mask for a brief DRIE step. By the latter, depths of hundreds of micrometers can be achieved with almost vertical sidewalls. Onto the etched surface, silica was sputtered for electrical insulation of the platinum layer, which served as a temperature measurement element and electrical resistance heater. The platinum was coated onto the surface by physical vapor deposition. The photoresist was then removed and Pyrex glass also bonded to the reverse side of the device.

2.6.2
Glass Micromachining

Glass microstructuring relies on similar cascaded fabrication procedures to those described above for silicon micromachining. Again, a practical example serves to demonstrate the general procedure applied [94]. Photosensitive glass was coated with a 1 nm chromium layer by sputtering and then with 1 μm photoresist by spin coating (see Figure 2.19). The photoresist was irradiated by a lithographic process. The glass was subsequently exposed to UV light to crystallize the non-irradiated parts which were protected by the chromium layer. The structured glass layer was capped with a bottom layer by fusion bonding at 550 °C and 1000 N m^{-2}. The crystallized glass parts were then etched with hydrofluoric acid. A catalyst coating was applied to the thus prepared microchannels.

2.6.3
Membrane Machining (for Hydrogen Separation)

Thin metallic palladium/silver membranes can be manufactured by cold rolling [95]. By several steps of rolling and annealing, 50 μm thick membranes are achieved. This rolling procedure increases the mechanical stability of the membranes, as expressed by the Vickers hardness. Cylindrical membrane separators were fabricated from the membranes by welding.

The generation of thin palladium membranes on ceramic surfaces needs more complicated approaches. Spray pyrolysis, chemical vapor deposition and sputtering have been applied. Electroless plating generates palladium particles from a solution containing amine complexes of palladium in the presence of reducing agents [96, 97]. Palladium nuclei need to be seeded on the surface prior to the coating procedure [97]. Ceramic surfaces such as α-alumina are first sensitized in acidic tin chloride and then palladium is seeded from acidic palladium ammonium chloride [97].

Osmosis effects can improve the plating procedure and repair defects [96]. A sodium chloride solution was fed through the tube side of the membrane, in order to induce diffusion of the water of the electroless plating solution through the membrane defects, to increase the palladium concentration locally and to heal the defects.

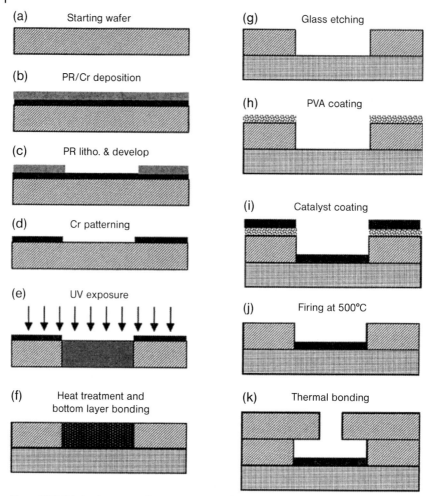

Figure 2.19 Fabrication steps of a MEMS-type methanol steam reformer by hydrofluoric acid etching [94].

MEMS technology was used to make a palladium membrane microreactor for hydrogen separation, with integrated devices for heating and temperature detection [98]. The reactor was composed of two channels separated by the membrane, which was composed of three layers. Two of these were deposited on a silicon wafer (see Figure 2.20), one made of silicon nitride by low-pressure chemical vapor deposition (0.3 μm thick) and one made of silicon oxide by temperature treatment (0.2 μm thick). These served as a perforated support for the palladium membrane. Etching of the wafer by potassium hydroxide then yielded one channel. Platinum/titanium films were deposited by electron beam, forming heaters and temperature sensors. A "blanket" of palladium was deposited onto one side of the support using

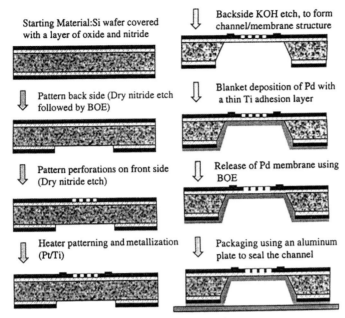

Figure 2.20 Microfabrication sequence for the silicon component of the palladium membrane reactor (by courtesy of Elsevier) [97].

a titanium film as adhesion layer. In this way a 700 μm wide and 17 mm long membrane of 200 nm thickness was realized. An aluminum plate sealed the permeate channel underneath the membrane. The second (retenate) channel was fabricated of polydimethylsiloxane applying a molding process.

A high mechanical stability of the membrane was demonstrated. Rupture was observed at more than 5 bar pressure, when the device was pressurized below the support structure on the permeate side. The pressure tolerance was much lower (1.4 bar), when the membrane was pressurized from the opposite direction. A pressure drop of 1 bar was even tolerated at a temperature of 500 °C when the higher pressure was put below the support structure.

The gas separation performance of the membrane was tested with a mixture of 90% nitrogen and 10% hydrogen at atmospheric pressure and a temperature of 500 °C. The permeate side of the membrane was evacuated. A hydrogen flux of $100 \, lm^{-2}s^{-1}$ was determined at 1 bar pressure drop.

A 750 nm thin palladium/silver membrane with 77 wt.% palladium was sputtered onto a silicon chip [99]. At 450 °C, a hydrogen flux of $22.4 \, lm^{-2}s^{-1}$ was determined. A 50 nm thin palladium membrane over a support structure with 5.5 μm holes withstood a pressure difference of 690 mbar without rupture [100]. A very thin (20 nm) palladium/silver membrane yielded a high $50 \, lm^{-2}s^{-1}$ hydrogen flux for a hydrogen partial pressure of only 35 mbar [101].

2.7
Catalyst-Coating Techniques

2.7.1
Surface Pre-treatment

To coat metallic surfaces with catalyst, a pre-treatment to improve the adherence is required [102]. Besides mechanical roughening, chemical and thermal pre-treatment are also applied frequently. Fecralloy, the construction material for metallic monoliths is usually pre-treated at temperatures between 900 and 1000 °C. An alumina layer about 1 μm thick is formed under these conditions on the Fecralloy surface, which is an ideal basis for catalyst coatings. However, metal oxide layers are formed on stainless steel and may also serve as adhesion layers.

Aluminum substrates are frequently pre-treated by anodic oxidation (see Figure 2.21) to generate a porous surface, which may serve as a catalyst support itself or as an adhesion layer for a catalyst support [102]. The surface area of the obtained alumina layer may reach $25 \, m^2 \, g^{-1}$ and a thickness of 70 μm and higher. The layer formed by anodic oxidation of aluminum and aluminum alloys is amorphous hydrated alumina. It is believed to contain boehmite, pseudoboehmite and physically adsorbed water [103]. It has the morphology of a packed array of hexagonal cells, containing a pore in the center. Thus the layer has a highly ordered porous structure and uniform thickness if an equally sized anode is positioned close to the oxidized surface.

2.7.2
Slurry/Sol Preparation

Once the surface is pre-treated, the coating slurry needs to be prepared. The most common method is to prepare a dispersion of finished catalyst, sometimes

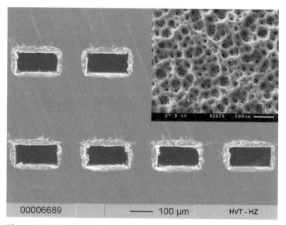

Figure 2.21 Porous alumina layer inside of microchannels in an aluminum microstructure device. The porous layer has been obtained by anodic oxidation (source: FZK).

including gelification steps [102]. Ceramic monoliths are usually wash-coated as follows. The catalyst carrier or the catalyst itself [104] is mixed with binder such as polyvinyl alcohol or methylhydroxyethyl cellulose [105], acid and solvent, usually water. Lower particle size improves adhesion [106, 107]. For ceramic monoliths, the particle size should be in the same range as the macro-pores of the ceramic support [107]. The carrier material, such as alumina, reacts with the acid, which increases the pH value of the slurry with time. This aging procedure may last more than 24 h [108]. Alumina powder only partially reacts with the acid and creates a gel.

It has been demonstrated that the slurry viscosity determines the thickness of the coating [109–112]. The viscosity itself is determined by the concentration of the particles, the pH value and the surfactant addition [109].

Additives to suppress particle agglomeration may be added to the suspension. This is crucial in the case of low particle size [106].

Sol–gel methods include a gelation procedure, also called peptization of the sol. The time needed for this procedure may vary considerably from hours to weeks. A sol is prepared by polycondensation of alkoxides [110].

Alumina sol may be prepared from aluminum alkoxide or pseudo-boehmite $AlO(OH) \cdot xH_2O$. Addition of additives such as urea provides the porosity of the catalyst layer by thermal decomposition during calcination [110].

The sol then serves as a binder for the particles which form the coating. Active metals can be incorporated into the sol. Usually sol–gel methods produce coatings of lower thickness, in the range of a few μm. Therefore hybrid methods between sol–gel and wash-coating are sometimes applied, making higher coating thickness possible. The sol–gel coating technique may be well suited for the coating of highly porous substrates, such as metal foams, because it generates uniform coatings, which are difficult to achieve by wash-coating in the case of foams.

2.7.3
Slurry/Sol Deposition

The most prominent coating technique for catalysts is wash-coating, which may be applied for ceramic and metallic monoliths as well as for plates of a stainless steel or aluminum plate heat-exchanger.

Materials that are routinely coated with catalyst wash-coats are ceramics such as cordierite, which is the construction material of ceramic monoliths, metals such as Fecralloy, the construction material of metallic monoliths, and stainless steel [113]. The amount of catalyst material which can be coated onto a monolith ranges between 20 and 40 g m^{-2}, while plate heat-exchangers may take up even more catalyst when coated prior to the sealing procedure because the access to the channels is better.

Wash-coating is frequently performed by dipping the substrate into the slurry or sol, which is the usual procedure for ceramic automotive exhaust gas treatment monoliths. The slurry excess is then removed by blowing off with air. An alternative is spray-coating, which requires a reduction in the viscosity of the slurry or sol [102]. Spray-coating has been used to improve the coating distribution on Fecralloy fibers [109]. Flame spray deposition was used for coating microchannels

[114]. The organo-metallic precursor compounds are guided through a flame and then driven toward the surface of the cooled substrate by thermophoresis. Spin coating was also used for wafer substrates.

Electrophoretic deposition is a colloidal process used to coat either aluminum or stainless steel [102, 115, 116]. Either adhesion layers or complete coatings may be achieved. To date, mostly alumina suspensions have been used for electrophoretic deposition. Microchannels have been coated with alumina nanoparticles dissolved in oxalic acid and mixed with alumina gel or glycerol [115]. 2 to 4 μm thick adhesion layers were formed.

Electrochemical deposition, also named electroplating, generates metallic or metal oxide coatings from their metal salts [102]. It may be used where metallic surfaces are required as catalyst or to generate metal oxide layers such as zirconia and lanthana. Electroless plating uses redox reactions to deposit metals. Electroless plating is also used to generate metallic membranes for hydrogen separation (see Section 2.6.3).

2.7.4
Temperature Treatment

After the deposition, drying and calcination steps usually follow, the latter being a temperature treatment in air or other gases for a defined duration. Calcination may include temperature ramps or not. However, the dried samples are not normally put immediately into a hot furnace but are heated up gradually. The final temperature of calcination needs to ensure that organic materials such as binders are removed completely.

Then, if only a catalyst carrier has been coated onto the surface and not a ready-made catalyst, the active species of the catalyst need to be impregnated onto the carrier.

A second drying and calcination procedure finalizes the catalyst preparation when an impregnation is performed. The drying procedure is crucial for the homogeneity of the active metals on the catalyst carrier [117]. Microwave and room-temperature drying result in more homogeneous distribution of active species such as nickel on cordierite monoliths compared to oven drying at 100 °C. Surface enrichment was observed with all drying procedures.

2.7.5
Alternative Coating Techniques

Thin layers of catalyst may be deposited onto the surface of silicon microchannels by physical vapor deposition. Silicon is the preferred material because the equipment for physical vapor deposition is available at microelectronics fabrication sites, which could also produce silicon microreactors. Physical vapor deposition such as cathodic sputtering, electron beam evaporation and pulsed laser deposition and also chemical vapor deposition create uniform metal surfaces with thicknesses in the nm range. Such coatings are rarely suitable as catalysts. However, a few exceptions such as hydrogen oxidation and reactions at very high temperature may exist.

Ceramic foams can be inserted into microdevices made from metals and polymers to enhance the surface area, act as catalyst supports or even work as heaters. Details of these processes can be found in Refs [50–63].

2.7.6
Automated Catalyst Coating

Apart from the commonly applied dip-coating process described above, the Metreon process was developed for coating unstructured Fecralloy metal foils [118]. The foils, in coils, were processed through rollers, where they received a corrugated structure. A heat treatment followed to generate the alumina layer (see Section 2.7.4). The foils were then wash-coated with the catalyst up to four times. Each coating step included drying and calcination. Finally the metal foil was recoiled. The coatings prepared by this method were highly resistant to thermal shocks.

Continuous wash-coating of micro-structured metal foils is possible using coating machines as shown in Figure 2.22. A continuously processed microstructured steel coil structured by wet-chemical etching is unwound from a roll and coated with the catalyst slurry. If required, the metal foil can be reduced to any plate size by laser-cutting at defined positions afterwards.

Acknowledgements

The authors would like to thank Tobias Hang from IMM for handling the copyright requests.

Figure 2.22 Semi-automated continuous coating machine for catalyst coating. The procedure is demonstrated here using a micro-structured metal foil unwound from a roll. The wash-coat slurry is fed through a slot die onto the metal foil. After coating the foil is sent through a drying compartment, an infrared calciner (not shown here) and wound up again (source: IMM).

References

1 Madou, M. (1997) *Fundamentals of Microfabrication*, CRC Press, London.

2 Menz, W. and Mohr, J. (1997) *Mikrosystemtechnik für Ingenieure*, Wiley-VCH, Weinheim, Germany.

3 Eigler, H. and Beyer, W. (1996) *Moderne Produktionsprozesse der Elektrotechnik, Elektronik und Mikrosystemtechnik*, Expert Verlag, Renningen, Germany.

4 Brandner, J.J. *et al.* (2006) Microfabrication in metals and polymers, in *Advanced Micro and Nanosystems Vol. 5: Micro Process Engineering* (eds H. Baltes, O. Brand, G.K. Fedder, C. Hierold, J. Korvink and O. Tabata), Wiley-VCH, Weinheim, Germany, pp. 267–320.

5 Brandner, J.J., Bohn, L., Schygulla, U., Wenka, A. and Schubert, K. (2003) Microstructure devices for thermal and chemical process engineering, in *Microreactors: Epoch-Making Technology for Synthesis* (ed. J.I. Yoshida), MCPT, 2001, J, CMC Publ. Co., Tokyo, pp. 75–87, 213–223.

6 Knitter, R., Dietrich, Th. (2006) Microfabrication in ceramic and glass, in *Advanced Micro & Nanosystems Vol. 5: Micro Process Engineering* (eds H. Baltes, O. Brand, G.K. Fedder, C. Hierold, J. Korvink and O. Tabata), Wiley-VCH Verlag GmbH, Weinheim, Germany.

7 Avila, P., Montes, M. and Miro, E.E. (2005) Monolithic reactors for environmental applications. A review on preparation techniques. *Chemical Engineering Journal*, **109**, 11–36.

8 Tomasic, V. and Jovic, F. (2006) State-of-the-art in the monolithic catalysts/reactors. *Applied Catalysis A: General*, **311**, 112–21.

9 Giroux, T., Hwang, S., Liu, Y., Ruettinger, W. and Shore, L. (2005) Monolithic structures as alternatives to particulate catalysts for the reforming of hydrocarbons for hydrogen generation. *Applied Catalysis B: Environmental*, **56**, 95–110.

10 Twigg, M.V. and Richardson, J.T. (1995) Scientific Bases for the preparation of heterogeneous catalysts, in *Preparation of Catalysts VI* (ed. G. Poncelet), Elsevier Ltd., pp. 345–59.

11 Dudfield, C.D., Chen, R. and Adcock, P.L. (2000) A compact CO selective oxidation reactor for solid polymer fuel cells powered vehicle applications. *Journal of Power Sources*, **86**, 214–22.

12 Takeda, T., Kunitomi, K., Horie, T. and Iwata, K. (1997) Feasibility study on the applicability of a diffusion-welded compact intermediate heat-exchanger to next-generation high temperature gas-cooled reactor. *Nuclear Engineering and Design*, **168**, 11–21.

13 Yu, X., Tu, S.-T., Wang, Z. and Qi, Y. (2006) Development of a microchannel reactor concerning steam reforming of methanol. *Chemical Engineering Journal*, **116**, 123–32.

14 Bier, W., Keller, W., Linder, G., Seidel, D., Schubert, K. and Martin, H. (1993) Gas-to-gas heat transfer in micro heat-exchangers. *Chemical Engineering Progress*, **32** (1), 33–43.

15 Schubert, K., Brandner, J., Fichtner, M., Linder, G., Schygulla, U. and Wenka, A. (2001) Microstructure devices for applications in thermal and chemical process engineering. *Nanoscale and Microscale Thermophysical Engineering*, **5**, 17–39.

16 Schuessler, M., Portscher, M. and Limbeck, U. (2003) Monolithic integrated fuel processor for the conversion of liquid methanol. *Catalysis Today*, **79–80**, 511–20.

17 Slocum, A.H. and Machine, P. (1992) Design: macromachine design philosophy and its applicability to the design of micromachines, Proc. IEEE MEMS 1992, Travemünde, Germany.

18 Boothroyd, G. and Knight, W.A. (1989) *Fundamentals of Machining and Machine Tools*, Marcel Dekker, Inc., New York, USA.

19 Evans, C. (1989) *Precision Engineering: An Evolutionary View*, Cranfield Press, Cranfield, Bedford, UK.

20 Snoeys, R. (1986) *Non-Conventional Machining Techniques, The State of the Art, Advances in Non-Traditional Machining,*

Proc. ASME Winter Annual Meeting, Anaheim, CA, USA.

21 Shaw, M.C. (1984) *Metal Cutting Principles*, Clarendon Press, Oxford, UK.

22 DeVries, W.R. (1992) *Analysis of Material Removal Processes*, Springer, New York, USA.

23 Chryssolouris, G. (1991) *Laser Machining*, Springer, New York, USA.

24 Reynaerts, P.-H.'s Heeren, D., Van Brussel, H., Beuret, C., Larsson, O. and Bertholds, A. (1997) Microstructuring of silicon by electro-discharge machining (EDM) – part II: applications. *Sensors and Actuators A*, **61**, 379–86.

25 Ho, K.H. and Newman, S.T. (2003) State of the art electrical discharge machining (EDM). *International Journal of Machine Tools and Manufacture*, **43** (13), 1287–300.

26 Petzow, G. (1994) *Metallographisches, Keramographisches und Plastographisches Ätzen*, Gebrüder Bornträger, Berlin, Germany.

27 Harris, T.W. (1976) *Chemical Milling*, Clarendon Press, Oxford, UK.

28 Drost, M.K., Wegeng, R.S., Martin, P. M., Brooks, K.P., Martin, J.L. and Call, C. (2000) MicroHeater, Proc. 4th International Conference on Microreraction Technology, AIChE, March 5–9, Atlanta, GA, USA, pp. 308–13.

29 Matson, D.W., Martin, P.M., Tonkovich, A.Y. and Roberts, G.L. (1998) Fabrication of a stainless steel microchannel microcombustor using a lamination process. *Proceedings of the SPIE*, **3514**, 286–392.

30 US 5,611,214 (March 18, 1997), US 8,811,062 (September 22, 1998).

31 Holladay, J.D., Brooks, K.P., Wegeng, R., Hu, J., Sanders, J. and Baird, S. (2007) Microreactor development for Martian in situ propellant production. *Catalysis Today*, **120** (1), 35–44.

32 Vansteenkiste, G., Boudeau, N., Leclerc, H., Barriere, T., Celin, J.C., Carmes, C., Roques, N., Millot, C., Benoit, C. and Boilat, C. (2004) Investigations in direct tooling for micro-technology with SLS, Proc. LANE 2004, Erlangen, Germany, pp. 425–34.

33 Fischer, P., Blatter, A., Romano, V. and Weber, H.P. (2004) Highly precise pulsed selective laser sintering of metal powders. *Laser Physics Letters*, 1–8.

34 Brandner, J.J., Hansjosten, E., Anurjew, E., Pfleging, W. and Schubert, K. (2007) Microstructure devices generation by selective laser melting, SPIE Photonics West, January 25–7, San Jose, CA, USA.

35 Boljanovic, V. (2004) *Sheet Metal Forming Processes and Die Design*, Industrial Press Inc., New York.

36 Shiu, P.P., Knopf, G.K., Ostojic, M. and Nikumb, S. (2008) Rapid fabrication of tooling for microfluidic devices via laser micromachining and hot embossing. *Journal of Micromechanics and Microengineering*, **18**, 025012.

37 Theis, H.E. (1999) *Handbook of Metalforming Processes*, CRC Press, Boca Raton.

38 Reed-Hill, R. (1991) *Physical Metallurgy Principles*, 3rd edn, PWS Publishing, Boston.

39 Kononenko, T.V., Garnov, S.V., Klimentov, S.M., Konov, V.I., Loubnin, E.N., Dausinger, F., Raiber, A. and Taut, C. (1997) Laser ablation of metals and ceramics in picosecond–nanosecond pulse width in the presence of different ambient atmospheres. *Applied Surface Science*, **109–110** (1), 48–51.

40 Momma, C., Nolte, S., Chichkov, B.N., Alvensleben, F.v. and Tünnermann, A. (1997) Precise laser ablation with ultrashort pulses. *Applied Surface Science*, **109–110** (1), 15–19.

41 Wang, X., Zhu, J., Bau, H. and Gorte, R.J. (2001) Fabrication of micro-reactors using tape-casting methods. *Catalysis Letters*, **77**, 173–7.4

42 Shin, Y., Kim, O., Hong, J.-C., Oh, J.-H., Kim, W.-J., Haam, S. and Chung, C.-H. (2006) The development of micro-fuel processor using low temperature co-fired ceramic (LTCC). *International Journal of Hydrogen Energy*, **31**, 1925–33.

43 Ehrfeld, W., Gärtner, C., Golbig, K., Hessel, V., Konrad, R., Löwe, H., Richter, T. and Schulz, C. (1997) Fabrication of components and systems for chemical and biological microreactors, in *Microreaction Technology, Proc. of The 1st*

Int. Conf. on Microreaction Technology (ed. W. Ehrfeld), Springer, Berlin, pp. 72–90.

44 Kolb, G., Cominos, V., Drese, K., Hessel, V., Hofmann, C., Löwe, H., Wörz, O. and Zapf, R. (2002) A novel catalyst testing microreactor for heterogeneous gas ohase reactions, in *Proc. 6th International Conference on Microreraction Technology* (eds P. Baselt, U. Eul, R.S. Wegeng, I. Rinard and B. Hoch), AIChE Topical Conference Proceedings, March 10–14, New Orleans, LA, USA, pp. 61–72.

45 Ziogas, A., Löwe, H., Küpper, M. and Ehrfeld, W. (2000) Electrochemical microreactors: a new approach for microreaction technology, in *Microreaction Technology: Proc. of the 3rd International Conference on Microreaction Technology* (ed. W. Ehrfeld), Springer, Berlin, Germany, pp. 136–50.

46 Meyer, H., Crämer, K., Kurtz, O., Herber, R., Friz, W., Schwiekendick, C., Ringtunatus, O. and Madry, C. (2002) Patent Application DE 10251658 A1.

47 Pfeifer, P., Görke, O., Schubert, K., Martin, D., Herz, S., Horn, U. and Gräbener, Th. (2005) Micromotive – development and fabrication of miniaturised components for gas generation in fuel cell systems, Proc. of the 8th Int. Conf. on Microreaction Techn. IMRET 8, April 10–4, Atlanta, GA, USA.

48 Paul, B.K., Hasan, H., Dewey, T., Alman, D. and Wilson, R.D. (2002) Development of aluminide microchannel arrays for high-temperature microreactors and micro-scale heat-exchangers, in *Proc. 6th International Conference on Microreraction Technology* (eds P. Baselt, U. Eul, R.S. Wegeng, I. Rinard and B. Hoch), AIChE Topical Conference Proceedings, March 10–14, New Orleans, LA, USA, pp. 202–11.

49 Bier, W., Keller, W., Linder, G., Seidel, D. and Schubert, K. (1990) Manufacturing and testing of compact micro heat-exchangers with high volumetric heat transfer coefficients, in *Symposium Volume*, DSC-Vol. 19, ASME, New York, pp. 189–97.

50 Heule, M., Vuillemin, S. and Gauckler, L.J. (2003) Powder-based ceramic meso and micro scale fabrication processes. *Advanced Engineering Materials*, **15**, 1237–45.

51 Yu, Z.Y., Rakurjar, K.P. and Tandon, A. (2004) Study of 3D micro-ultrasonic machining. *Journal of Manufacturing Science Engineering, Trans ASME*, **126**, 727–32.

52 Knitter, R., Günther, E., Maciejewski, U. and Odemer, C. (1994) Preparation of ceramic microstructures, cfi/Ber. *Deutsche Keramische Gesellschaft*, **71**, 549–56.

53 Mutsuddy, B.C. and Ford, R.G. (1995) *Ceramic Injection Molding*, Chapman & Hall, London, GB.

54 Griffith, M.L. and Halloran, J.W. (1996) Freeform fabrication of ceramics via stereo lithography. *Journal of the American Ceramic Society*, **79**, 2601–8.

55 Blazdell, P.F., Evans, J.R.G., Edirisinghe, M.J., Shaw, P. and Binstead, M.J. (1995) The computer aided manufacture of ceramics using multiplayer jet printing. *Journal of Materials Science Letters*, **14**, 1562–5.

56 Agrarwala, M.K., Bandyopadhyay, A., van Weeren, R., Safari, A., Danforth, S.C., Langrana, N., Jamalabad, V.R. and Whalen, P.J. (1996) FDC, rapid fabrication of structural component. *American Ceramic Society Bulletin*, **75**, 60–5.

57 Evans, J.R.G. (1996) Injection molding, in *Materials Science and Technology Vol. 17a, Processing of Ceramics Part 1* (ed. R.J. Brook), Wiley-VCH, Weinheim, Germany, pp. 267–311.

58 Bauer, W. and Knitter, R. (2002) Development of a rapid prototyping process chain for the production of ceramic microcomponents. *Journal of Materials Science*, **37**, 3127–40.

59 Mistler, R.E. (1995) The principles of tape casting and tape casting applications, in *Ceramic Processing* (eds R.A. Terpstra, P.P.A.C. Pex and A.H. de Vries), Chapman & Hall, London, UK.

60 Ritzhaupt-Kleissl, H.-J., von Both, H., Dauscher, M. and Knitter, R. (2005) Further ceramic replication techniques, in *Advanced Micro and Nanosystems Vol.*

4, Microengineering of Metals and Ceramics (eds H. Baltes, O. Brand, G.K. Fedder, C. Hierold, J. Korvink and O. Tabata), Wiley-VCH, Weinheim, Germany.

61 Su, B., Button, T.W., Schneider, A., Singleton, L. and Prewett, P. (2002) Embossing of 3D ceramic microstructures. *Microsystems Technologies*, **8**, 359–62.

62 Meschke, F., Riebler, G., Hessel, V., Schürer, J. and Baier, T. (2005) Hermetic gas-tight ceramic microreactors. *Chemical Engineering and Technology*, **8**, 465–73.

63 Haas-Santo, K., Görke, O., Pfeifer, P. and Schubert, K. (2002) Catalyst coatings for microstructure reactors. *Chimia*, **56**, 605–10.

64 Giselbrecht, S., Gottwald, E., Schlingloff, G., Schober, A., Truckenmüller, R., Weibezahn, K.F. and Welle, A. (2005) Highly adaptable microstructured 3D cell culture platform in the 96 well format for stem cell differentiation and characterization, Proc. 9th Int. Conf. on Miniaturized Systems for Chemistry and Life Sciences µTAS 2005, October 9–13, (2005), Boston, MA, USA, to be published.

65 Ehrenstein, G.W. and Erhard, G. (1983) *Konstruieren mit Polymerwerkstoffen ein Bericht zum Stand der Technik*, Hanser-Verlag, München, Germany.

66 Mohr, J.A., Last, A., Hollenbach, U., Oka, T. and Wallrabe, U. (2003) A modular fabrication concept for microoptical systems. *Journal of Lightwave Technolgy*, **21**, 643–7.

67 Ruprecht, R., Benzler, T., Holzer, P., Müller, K., Norajitra, P., Piotter, V. and Ulrich, H. (1999) Spritzgießen von Mikroteilen aus Kunststoff, Metall und Keramik. *Galvanotechnik*, **90**, 2260–7.

68 Heckele, M. and Schomburg, W.K. (2004) Review on micro molding of thermoplastic polymers. *Journal of Micromechanics and Microengineering*, **14**, R1–14.

69 Ehrfeld, W., Hessel, V. and Löwe, H. (2000) *Microreactors*, Wiley-VCH, Weinheim.

70 Becker, H. and Heim, U. (2000) Hot embossing as a method for the fabrication of polymer high aspect ratio structures. *Sensors Actuators A*, **83**, 130–5.

71 Shiu, P.P., Knopf, G.K., Ostojic, M. and Nikumb, S. (2008) Rapid fabrication of tooling for microfluidic devices via laser micromachining and hot embossing. *Journal of Micromechanics and Microengineering*, **18**, 025012.

72 Leech, P.W. (2008) Pattern replication in polypropylene films by hot embossing. *Microelectronic Engineering*, **85** (1), 181–6.

73 Pettit, G.H. and Sauerbrey, R. (1993) Pulsed ultraviolet laser ablation. *Applied Physics A*, **56**, 51–63.

74 Pfleging, W., Hanemann, T., Bernauer, W. and Torge, M. (2001) Laser micromachining of mold inserts for replication techniques – State of the art and application. *Proceedings of the SPIE*, **4274**, 331–45.

75 Cheng, J.-Y., Wie, C.-W., Hsu, K.-H. and Young, T.-H. (2004) Direct-write laser micromachining and universal surface modification of PMMA for device development. *Sensors and Actuators B*, **99**, 186–96.

76 Gebharth, A. (1996) *Rapid Prototyping*, Hansa Verlag, München, Germany.

77 Ikuta, K., Hirowatari, K. and Ogata, T. (1994) Three dimensional integrated fluid systems (MIFS) fabricated by stereo lithography, Proc. IEEE Int. Workshop on Micro Electro Mechanical Systems, MEMS'94, Osio, Japan, pp. 1–6.

78 Ikuta, K., Hasegawa, T., Adachi, T. and Maruo, S. (2000) Micro-stereolithography and its application to biomedical IC-chip, Proc. IEEE Int. Workshop on Micro Electro Mechanical Systems, MEMS'2000, 739ff.

79 Ikuta, K. (2003) Module micro chemical device based on biochemical IC chips – From 3D micro/nano fabrication toward biomedical applications, Proc. 1st Int. Workshop on Micro Chemical Plant Techn., Feb 3–4, Kyoto, J, pp. 54–65.

80 Bacher, W. and Saile, V. (2003) LIGA and AMANDA technologies for the fabrication of advanced micro-devices,

Proc. 2003 JSME-IPP/ASME-ISPS Joint Conf. on Micromechatronics for Information and Precision Equipment, Yokohama, J, June 16–18, pp. 133–7.

81 Truckenmüller, R., Ahrens, R., Bahrs, H., Cheng, Y., Fischer, G. and Lehmann, J. (2005) Micro ultrasonic welding – joining of chemically inert, high temperature polymer microparts for single material fluidic components and systems, Proc. of DTIP, June 1-3, Montreux, CH, to be published.

82 Bader, R., Jacob, P., Volk, P. and Moritz, H. (1999) Process for joining microstructured plastic parts and component produced by this process, European Patent No. WO 99/25783.

83 Bachmann, F. and Russek, U. (2002) Laser welding of polymers using high power diode lasers. *Proceedings of the SPIE*, **4637**, 505–18.

84 Sato, K., Kurosaki, Y., Saito, T. and Satoh, I. (2002) Laser welding of plastics transparent to near-infrared radiation. *Proceedings of the SPIE*, **4637**, 528–36.

85 www.clearweld.com (accessed 31.10.08).

86 Klotzbuecher, T., Letschert, M., Braune, T., Drese, K.-S. and Doll, T. (2006) Diode laser welding for packaging of transparent micro structured polymer chips. *Proceedings of the SPIE*, **6107**, 34.

87 Pfleging, W., Baldus, O., Bruns, M., Baldini, A. and Bemporad, E. (2005) Laser assisted welding of transparent polymers for micro chemical engineering and life science, Proc. Int. Conf. "Laser and Applications in Science and Technology (LASE)", Photonic West 2005, San Jose, CA, USA, Jan 22–7, to be published in Proc. SPIE Vol. 5713.

88 Hessel, V. and Löwe, H. (2002) Mikroverfahrenstechnik: Komponenten – Anlagenkonzeptionen – Anwenderakzeptanz. *Chemie Ingenieur Technik*, **74** (1–2), 17–30, and (3), 185–207, and (4), 381–400.

89 Arana, L.R. (2003) High-Temperature Microfluidic Systems for Thermally-Efficient Fuel Processing, PhD thesis, Massachusetts Institute of Technology.

90 Arana, L.R., Baertsch, C.D., Schmidt, R.C., Schmidt, M.A. and Jensen, K.F. (2003) Combustion – assisted hydrogen production in a high-temperature chemical reactor/heat-exchanger for portable fuel cell applications, 12th International Conference on Solid-State Sensors, Actuators, and Microsystems (Transducers 03), Boston, MA.

91 Arana, L.R., Schaevitz, S.B., Franz, A.J. and Jensen, K.F. (2002) A microfabricated suspended-tube chemical reactor for fuel processing, Proceedings of the 15th IEEE International Conference on Micro ElectroMechanical Systems, IEEE, New York, pp. 212–15.

92 Bustillo, J.M., Howe, R.T. and Muller, R.S. (1998) Surface micromachining for microelectromechanical systems. *Proceedings of the IEEE*, **86**, 1552–74.

93 Pattekar, A.V. and Kothare, M.V. (2004) A microreactor for hydrogen production in micro fuel cells. *Journal of Microelectromechanical Systems*, **13**, 7–18.1

94 Kim, T. and Kwon, S. (2006) Design, fabrication and testing of a catalytic microreactor for hydrogen production. *Journal of Micromechanics and Microengineering*, **16**, 1760–8.

95 Tosti, S., Bettinali, L. and Violante, V. (2000) Rolled thin Pd and Pd-Ag membranes for hydrogen separation and production. *International Journal of Hydrogen Energy*, **25**, 319–25.

96 Gepert, V., Kilgus, M., Schiestel, T., Brunner, H., Eigenberger, G. and Merten, C. (2006) Ceramic supported capillary Pd membranes for hydrogen separation: potential and present limitations. *Fuel Cells*, **6**, 472–81.

97 Li, A., Liang, W. and Hughes, R. (2000) Fabrication of dense palladium composite membranes for hydrogen separation. *Catalysis Today*, **56**, 45–51.

98 Franz, A.J., Jensen, K.J. and Schmidt, M.A. (2000) Palladium membrane microreactors, in *Microreaction Technology: 3rd International Conference on Microreaction Technology, Proc of IMRET 3* (ed. W. Ehrfeld), Springer-Verlag, Berlin, pp. 267–76.

99 Gielens, E.C., Tong, H.D., van Rijn, C.J.M., Vorstman, M.A.G. and Keurentjes, J.T.F. (2002) High flux palladium-silver alloy membranes fabricated by microsystem technology. *Desalination*, **147**, 417–23.

100 Karnik, S.V., Hatalis, M.K. and Kothare, M.V. (2001) Palladium based micro-membrane for water gas shift reaction and hydrogen gas separation, in *Microreaction Technology – IMRET 5: Proc. of the 5th International Conference on Microreaction Technology* (eds S.V. Karnik, M.K. Hatalis and M.V. Kothare), Springer-Verlag, Berlin, pp. 295–302.

101 Wilhite, B.A., Weiss, S.E., Ying, J.Y., Schmidt, M.A. and Jensen, K.F. (2006) High-purity hydrogen generation in a microfabricated 23 wt.% Ag-Pd membrane device integrated with $8:1$ $LaNi_{0.95}Co_{0.05}O_3/Al_2O_3$ catalyst. *Advanced Materials*, **18**, 1701–4.

102 Meille, V. (2006) Review of methods to deposit catalysts on structured surfaces. *Applied Catalysis A: General*, **315**, 1–17.

103 Ganley, J.C., Riechmann, K.L., Seebauer, E.G. and Masel, R.I. (2004) Porous anodic alumina optimized as a catalyst support for microreactors. *Journal of Catalysis*, **227**, 26–32.

104 Zapf, R., Becker-Willinger, C., Berresheim, K., Holz, H., Gnaser, H., Hessel, V., Kolb, G., Löb, P., Pannwitt, A.-K. and Ziogas, A. (2003) Detailed characterization of various porous alumina based catalyst coatings within microchannels and their testing for methanol steam reforming. *Chemical Engineering Research and Design A*, **81**, 721–9.

105 Germani, G., Stefanescu, A., Schuurman, Y. and van Veen, A.C. (2007) Preparation and characterization of porous alumina-based catalyst coatings in microchannels. *Chemical Engineering Science*, **62**, 5084–91.

106 Agrafiotis, C. and Tsetsekou, A. (2000) The effect of powder characteristics on washcoat quality. Part I: alumina washcoats. *Journal of the European Ceramic Society*, **20**, 815–24.

107 Avila, P., Montes, M. and Miro, E.E. (2005) Monolithic reactors for environmental applications. A review on preparation techniques. *Chemical Engineering Journal*, **109**, 11–36.

108 Cristiani, C., Valentini, M., Merazzi, M., Neglia, S. and Forzatti, P. (2005) Effect of aging time on chemical and rheological evolution in g-Al_2O_3 slurries for dip-coating. *Catalysis Today*, **105**, 492–8.

109 Meille, V., Pallier, S., Santa Cruz Bustamante, G.V., Roumanie, M. and Reymond, J.-P. (2005) Deposition of g-Al_2O_3 layers on structured supports for the design of new catalytic reactors. *Applied Catalysis A: General*, **286**, 232–8.

110 Tomasic, V. and Jovic, F. (2006) State-of-the-art in the monolithic catalysts/reactors. *Applied Catalysis A: General*, **311**, 112–21.

111 Hwang, S.-M., Kwon, O.J. and Kim, J.J. (2007) Method of catalyst coating in micro-reactors for methanol steam reforming. *Applied Catalysis A: General*, **316**, 83–9.

112 Haas-Santo, K., Fichtner, M. and Schubert, K. (2001) Preparation of microstructure compatible porous supports by sol-gel synthesis for catalyst coatings. *Applied Catalysis A: General*, **220**, 79–92.

113 Giroux, T., Hwang, S., Liu, Y., Ruettinger, W. and Shore, L. (2005) Monolithic structures as alternatives to particulate catalysts for the reforming of hydrocarbons for hydrogen generation. *Applied Catalysis B: Environmental*, **56**, 95–110.

114 Thybo, S., Jensen, S., Johansen, J., Johannesen, T., Hansen, O. and Quaade, U.J. (2004) Flame spry deposition of porous catalysts on surfaces and in microsystems. *Journal of Catalysis*, **223**, 271–7.

115 Wunsch, R., Fichtner, M., Görke, O., Haas-Santo, K. and Schubert, K. (2002) Process of applying Al_2O_3 coatings in microchannels of completely manufactured microstructured reactors. *Chemical Engineering and Technology*, **25** (7), 700–3.

116 Pfeifer, P., Görke, O. and Schubert, K. (2002) Washcoats and electrophoresis with coated and uncoated nanoparticles on microstructured metal foils and microstructured reactors, Proceedings of the 6th International Conference on Microreaction Technology, IMRET 6, March 11–14, AIChE Pub. No. 164, New Orleans, USA, pp. 281–5.

117 Villegas, L., Masset, F. and Guilhaume, N. (2007) Wet impregnation of alumina-washcoated monoliths: effects of the drying procedure on Ni distribution an on autothermal reforming activity. *Applied Catalysis A: General*, **320**, 43–55.

118 Lylykanas, R. and Lappi, P. (1991) SAE Paper 910614.

3
Fuel Processing for Hydrogen Generation

Kishori Deshpande

3.1
Introduction

Hydrogen-based fuel cells are attractive owing to their high efficiency and benign by-products. Hydrogen, however, does not occur naturally as a gaseous fuel and hence has to be tapped from other sources. Among the possible options such as electrolysis and hydrocarbon reforming, the latter option is commonly used to produce hydrogen. Reforming can either be carried out on board or in a separate unit. The latter option entails storing hydrogen for on-demand usage. While hydrogen generation using various primary fossil fuels has been explored for large scale use, its adaptation is not feasible for micropower generation owing to design complexities The focus of this chapter is to discuss the various processes employed for hydrogen generation for power systems of the order of a few watts. Further, issues related to hydrogen storage and purification for microsystems are also addressed.

Fuel reforming is the conversion of the primary feedstock to the desired fuel gas composition as required by the fuel cell system (Table 3.1) [1]. For portable power generation, proton exchange membrane (PEM) fuel cells have been proposed as battery replacements or in conjunction with batteries to enhance the power density [2]. Thus, based on the fuel cell selection and specific fuel requirement for micropower generation, the fuel processing design should yield CO free hydrogen. Further, to avoid fuel cell catalyst poisoning, it should also be sulfur-free.

As noted previously, hydrogen is generated from a variety of feedstocks including petroleum, coal, natural gas and biofuels for large scale power production. Natural gas and petroleum liquids contain organic sulfur compounds which should be eliminated prior to fuel processing. For portable power application this implies designing a compact desulfurization unit. Further, lower operating temperatures and pressures are favored to address systems integration issues. For example, high-temperature operation might pose problems in finding materials with similar thermal expansion coefficients or sealants for a leak-proof seal. Hence most of the microreactors are designed for lower hydrocarbons such as methanol

Microfabricated Power Generation Devices. Edited by Alexander Mitsos and Paul I. Barton
Copyright © 2009 WILEY-VCH Verlag GmbH & Co. KGaA, Weinheim
ISBN: 978-3-527-32081-3

Table 3.1 Fuel requirement for different fuel cells (Reproduced from Ref. [1] with permission).

Gas species	PEM	AFC	PAFC	MCFC	SOFC
H_2	Fuel	Fuel	Fuel	Fuel	Fuel
CO	Poison (>10 ppm)	Poison	Poison (>0.5%)	Fuel	Fuel
CH_4	Diluent	Diluent	Diluent	Diluent[a]	Diluent[a]
CO_2 and H_2O	Diluent	Poison	Diluent	Diluent	Diluent
S	Few studies	Unknown	Poison (>50 ppm)	Poison (>0.5 ppm)	Poison (>1 ppm)

a   In internal reforming fuel cells, acts as a fuel.

or propane owing to the design simplicity and operating conditions required for these fuels. Three basic fuel reforming techniques used for hydrogen generation include steam reforming, partial oxidation and autothermal reforming [3]. Depending on whether the fuel is reformed in a separate reactor integrated with the fuel cell or within the fuel cell anode chamber, indirect or direct internal reforming can be achieved, respectively. Pyrolysis and ammonia cracking can also be used to produce hydrogen. Further, the generated hydrogen should be cleaned of concomitant CO, either by preferential oxidation or membrane separation (Figure 3.1). The following section discusses different hydrogen generation approaches along with the associated fuel cleaning strategies.

3.2
Hydrogen Generation

3.2.1
Reforming Techniques

As mentioned in the previous section, hydrogen can be generated using different techniques such as electrolysis and hydrocarbon reforming. Among the various routes, hydrocarbon reforming is commonly practised for large scale hydrogen generation. However, for portable power generation it entails design modification for easier adaptation. In the section below different reforming processes are discussed in detail.

3.2.1.1   Steam Reforming
The basic endothermic reforming reactions for methanol and generic hydrocarbon are given below:

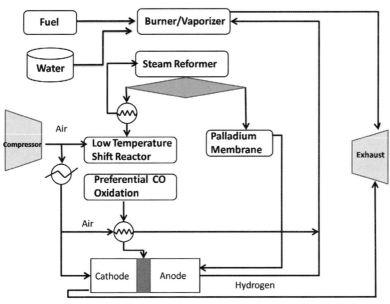

Figure 3.1 Schematic for fuel processing.

$$CH_3OH + H_2O \rightarrow CO_2 + 3H_2 \qquad\qquad \Delta H = 49.7\,\mathrm{kJ\,mol^{-1}} \qquad\qquad (1)$$

$$CO + H_2O \rightarrow CO_2 + H_2 \qquad\qquad \Delta H = -41\ \mathrm{kJ\,mol^{-1}} \qquad\qquad (2)$$

$$C_nH_m + nH_2O \rightarrow nCO + (m/2 + n)\,H_2 \qquad\qquad\qquad\qquad\qquad\qquad (3)$$

The reforming reaction and associated water gas shift reaction (Reaction (2)) are typically carried out over a supported catalyst and at elevated temperature, depending on the fuel involved. For methanol, the reaction is performed over Cu supported on alumina or Pd/Zn alloy catalysts at modest temperature 200 °C [4–10]. Pd/Zn-based alloys are preferred because Cu-based catalysts pose sintering problems at higher temperatures. Further, they also lead to lower CO formation [4]. Since Reactions (2) and (3) occur simultaneously, the product stream comprises hydrogen, CO, CO$_2$ and unreacted fuel feed. The exact composition is dictated by the operating temperature, pressure, composition of the feed gas and steam to fuel ratio. For example, in Reaction (1) the Le Chatelier principle indicates that lowering the operating pressure would favor the reaction to the right and hence higher hydrogen formation.

For fuels other than methanol, reforming is carried out using nickel-based catalysts at temperatures >500 °C [11, 12]. However, at higher temperatures carbon formation occurs through either pyrolysis (Reaction (4)) or the Boudouard reaction (Reaction (5)). One way to prevent carbon formation is introduce steam to promote water gas shift (Reaction (3)) and carbon gasification reactions (Reaction (6)).

Hence, for practical purposes, the steam to carbon ratio usually lies between 2 and 3 to avoid carbon formation. Coking can also be reduced by choosing a noble metal catalyst or adding magnesia or potassia to the catalyst support [2]. Water storage is one of the major disadvantages of steam reforming for microsystems since carrying water reduces the energy density. Further, recycling water is very challenging.

$$CH_4 \rightarrow C + 2H_2 \qquad \Delta H = 75\,kJ\,mol^{-1} \tag{4}$$

$$2CO \rightarrow C + CO_2 \tag{5}$$

$$C + H_2O \rightarrow CO + H_2 \tag{6}$$

3.2.1.2 Partial Oxidation

Another approach to hydrogen generation involves using oxygen instead of steam (Reaction (7)). Partial oxidation can be carried out with or without a catalyst. As one would expect, noncatalytic partial oxidation reactions occur at high temperature (>1000 °C). An advantage of this process is that a desulfurization unit is not required. However, the H_2S formed should be removed at a later stage. While these reactions have been used at large scale, they are difficult to scale down and control of the reaction is difficult due to its exothermic nature. On the other hand catalytic partial oxidation (CPO) occurs at lower temperature in the presence of platinum- or nickel-based catalysts [13, 14]. Partial oxidation yields less hydrogen than steam reforming. For example, compared to 3 moles of hydrogen generated in steam reforming, 1 mole of methanol yields 2 moles of hydrogen during partial oxidation. Further, if air is used as the oxidant, it leads to lowering of the partial pressure of hydrogen in the exit stream and thus a lower Nernst potential. A key advantage of the process, however, is the simplified system design due to the absence of steam.

$$CH_3OH + 0.5\,O_2 \rightarrow CO_2 + 2H_2 \quad \Delta H^\circ = -192.3\,kJ\,mol^{-1} \tag{7}$$

$$C_nH_m + 0.5n\,O_2 \rightarrow nCO + 0.5m\,H_2 \tag{8}$$

3.2.1.3 Autothermal Reforming

Autothermal reforming combines partial or total oxidation and steam reforming [15, 16]. Thus an oxidant (oxygen or air) and steam are simultaneously introduced, resulting in higher hydrogen than with partial oxidation but lower than with steam reforming. The advantage of the process is that no external heat is required since the energy required by steam reforming is supplied by the exothermic partial oxidation of the fuel.

$$2CH_3OH + 0.5\,O_2 + H_2O \rightarrow 2CO_2 + 5H_2 \tag{9}$$

$$2C_nH_m + n\,H_2O + 1.5n\,O_2 \rightarrow 2nCO_2 + (n + m)\,H_2 \tag{10}$$

3.2.1.4 Pyrolysis

Another alternative for generating hydrogen is the decomposition of hydrocarbons by heating in the absence of air or water, leading to pure hydrogen and carbon. Thus no carbon oxides are formed and, in the absence of subsequent clean-up units, the design becomes compact and modular.

$$C_m H_n \rightarrow mC + n/2\, H_2 \tag{11}$$

However, carbon management is critical for these reactors since it may foul the catalyst or block the reactor. Typically, carbon removal is achieved by flowing air over the bed to burn off carbon as carbon dioxide. Despite rendering the design compact owing to the reactions involved, not much work has been reported on the pyrolysis of hydrocarbons in microreactors. This might be explained by the associated carbon management issues.

3.2.1.5 Ammonia Cracking

Ammonia is considered a good hydrogen carrier because it can be easily liquefied, has high energy density, is relatively inexpensive and the by-products do not involve CO. The disadvantage of this fuel is its high toxicity and irreversible damage to the fuel cell performance even at 50 ppm concentration for a PEM fuel cell. Additionally, the reaction is highly endothermic. Nevertheless, ammonia reforming is carried out using supported Ru or Ni-based catalyst at 800–900 °C for higher ammonia conversion [17–19]. Since ammonia decomposes into nitrogen and hydrogen extensive clean-up is unnecessary except for removal of undecomposed ammonia from the exit stream.

$$2NH_3 \rightarrow N_2 + 3\,H_2 \quad \Delta H = 46.4 \, \text{kJ mol}^{-1} \tag{12}$$

3.3
Related Processes: Hydrogen Clean-Up

For on board hydrogen reformers involving hydrocarbons, CO removal is essential to prevent fuel cell catalyst poisoning. For example, a methanol reformer operating at 200 °C generates at least 0.1% CO, depending on pressure and water content [1]. The important problem of CO removal in the reformed stream is addressed using the water gas shift reaction (Reaction (2)).

Though the reaction occurs simultaneously with steam reforming, the thermodynamics of the reaction favor CO formation under the fuel reforming conditions. Hence, to reduce the CO content, the exit stream is cooled and passed over a catalyst bed which promotes the shift reaction. Iron–chromium catalysts promote shift reaction at high temperature (400–500 °C) followed by cooling the exit stream to 200 °C and second stage purification over copper-based catalyst. Even with the two-stage CO removal the exit stream may contain up to 5000 ppm. CO, which is about 100 times higher than the PEM requirement. To further reduce the poison content either of the following methods is used. In a selective oxidation reactor a small amount of air (2%) is introduced into the fuel stream over a precious metal catalyst bed [20–22]. The catalyst preferentially adsorbs CO which is oxidized. The reactor adds to the system complexity since carefully controlled quantities of air must be introduced to prevent explosive conditions. To avoid the danger of explosion, methanation reactors can be used wherein carbon monoxide reacts with hydrogen to form methane (Reaction (13)). However, this entails hydrogen consumption and hence lowering of fuel cell efficiency. Typically platinum- and ruthenium-based catalysts are used for these reactions.

$$CO + 3H_2 \rightarrow CH_4 + H_2O \quad \Delta H = -206\,kJ\,mol^{-1} \tag{13}$$

Palladium membranes which are selectively permeable to hydrogen offer an attractive solution for the CO reduction problem. In most cases palladium is alloyed with silver to overcome the problem of hydrogen embrittlement. Besides making the design compact these membranes yield high hydrogen selectivity and are chemically resistant to carbon dioxide and carbon monoxide [23–26]. A sweep stream or a pump is, however, needed for the separation.

3.4
Alternative Hydrogen Production Techniques

An alternative to onboard hydrogen generation involves reacting chemical hydrides such as LiH, NaH and MgH$_2$ with either water or alcohol [27, 28]. These reactions can be carried out at room temperature and the reaction products need to be recycled off board the vehicle. Table 3.2 gives the different types of chemical hydrides and their hydrogen content [3]. While this area has been investigated extensively, it is outside the scope of this chapter and is not discussed in detail.

Table 3.2 Hydrogen content for different hydrides (Reproduced from Ref. [3] with permission).

Name	Formula	Percent Hydrogen	Density (kg l^{-1})	Volume (l) to store 1 kg H$_2$
Lithium hydride	LiH	12.68	0.82	6.5
Beryllium hydride	BeH$_2$	18.28	0.67	8.2
Sodium hydride	NaH	4.3	0.92	25.9
Aluminum hydride	AlH$_3$	10.8	1.3	7.1
Potassium hydride	KH	2.51	1.47	27.1
Calcium hydride	CaH$_2$	5	1.9	1.1
Lithium borohydride	LiBH$_4$	18.51	0.67	8
Sodium borohydride	NaBH$_4$	10.58	1	9.5
Aluminum borohydride	Al(BH$_4$)$_3$	16.91	0.545	11
Palladium hydride	Pd$_2$H	0.47	10.78	20

3.5
Materials Selection

For microreactors, decrease in linear dimensions leads to enhanced heat and mass transfer. Since fuel reforming occurs at high temperature (>300 °C), proper materials selection is important for thermal management and systems integration. The higher operating temperature limits the choices to silicon, aluminum, stainless steel and high-temperature ceramics. Further, for endothermic reactions such as steam reforming, efficient heat transfer is important for the desired catalytic activity and prevention of undesirable side products. Thus the materials selection is specific to the operating conditions and reaction thermodynamics. In addition to the above-mentioned parameter, other issues include pressure drop along the reactor, sealing fluidic connections and system monitoring. For example, thermocouples used for temperature measurement at microscale might act as a heat sink and interfere with the process monitoring. Similarly, a slight mismatch in the thermal expansion coefficients of the sealing material at high temperatures may lead to microcracks and hence leaky seals.

In the following section different reforming examples are discussed with emphasis on reaction chemistry, materials selection and reactor performance.

3.6
Literature Examples

Numerous groups are working on on-board hydrogen generation using hydrocarbon reforming. While there are many references to their use for fuel cells, very few authors report complete reformer–fuel cell systems. Microreactors, owing to the decrease in linear dimensions, possess excellent heat and mass transfer. Hence, for reformers operating at high temperatures (>300 °C) thermal management becomes a key criterion. While there is exciting work on microreformers in academia and industry alike, owing to space constraints only a couple of examples are discussed in the light of hydrogen generation and purification along with their thermal efficiencies wherever reported.

3.6.1
Methanol Partial Oxidation

Wilhite et al. [13] report high-purity hydrogen generation using a palladium-based micromembrane device that integrates nanometer scale palladium silver film with complex oxide methanol reforming catalysts in a silicon microreactor. The authors fabricate the microreformer using bulk micromachining techniques such as photolithography and wet/dry etching. Finally, the device is packaged with another patterned silicon wafer using quick drying epoxy (Figure 3.2). The authors heat the device electrically to 400–47 °C for methanol partial oxidation using a LaNiCoO$_3$-based catalyst. The high working temperature justifies the specific materials

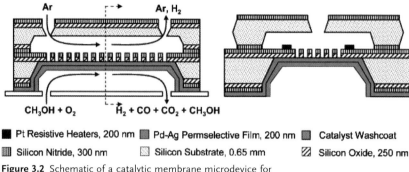

Ar   Ar, H₂

CH₃OH + O₂   H₂ + CO + CO₂ + CH₃OH

■ Pt Resistive Heaters, 200 nm ■ Pd-Ag Permselective Film, 200 nm ■ Catalyst Washcoat

▥ Silicon Nitride, 300 nm ▦ Silicon Substrate, 0.65 mm ▨ Silicon Oxide, 250 nm

Figure 3.2 Schematic of a catalytic membrane microdevice for high purity hydrogen generation (reproduced from Ref. [13] with permission).

selection that is silicon, in the study. As mentioned in the previous section, owing to the absence of steam the reaction scheme is simplified. Thermal management of their system is explained by localized heating of the reactor using a nitride–oxide electrical insulation layer. The insulation membrane allows placement of the resistive heaters directly on the reaction zone making heating of the entire microdevice unnecessary.

Subsequent purification of the generated hydrogen is carried out using a 0.2 micron Pd–Ag alloy membrane, resulting in purified hydrogen fluxes up to 2.2 mol m^{-2}s^{-1}. The authors report overall hydrogen selectivities greater than 90% with an extracted hydrogen selectivity of 47% at 475 °C and an oxygen to fuel ratio of 0.43, see Figure 3.3. The work highlights the integrated hydrogen reforming–purification unit approach for portable power generation.

3.6.1
Samsung Micro Fuel Processor

Another example of fuel reforming involves a silicon-based microvaporizer–reformer developed by Samsung for cell phone application [6]. The silicon device, consisting of a vaporizer chamber followed by the reactor, is microfabricated in two different configurations viz., serpentine Figure 3.4, and parallel channels using standard photolithography processes. The authors reform methanol by electrically heating the reactor to 260 °C using commercial Cu/ZnO supported on Al₂O₃ catalyst. The average hydrogen production was 0.0445 mol h^{-1} which translates to 2.4 W assuming 80% fuel cell efficiency. The exit stream, consisting of 1.5% CO, is not purified and the investigators propose using it directly in a high-temperature fuel cell. The work indicates that the serpentine channeled reactor gives better performance than parallel channels owing to higher residence times in the former case. Further, catalyst durability tests indicate that the shape of the reactor influences the life of the catalyst (Figure 3.5). The authors conclude that a combination of reformer and battery should be used for cell phone

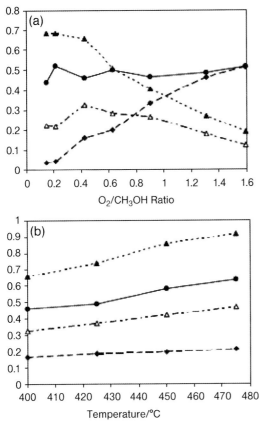

Figure 3.3 Results obtained for a 23 wt% Ag–Pd film washcoated with 1.69 mg of 8:2 LaNi$_{0.95}$Co$_{0.05}$O$_3$/Al$_2$O$_3$ catalyst. (a) At a constant membrane temperature of 400 °C and varying O$_2$/CH$_3$OH ratio and (b) for a varying membrane temperature at a constant O$_2$/CH$_3$OH ratio of 0.43. Total methanol conversion (●), total hydrogen selectivity (▲), extracted hydrogen selectivity (△), and selectivity towards CO$_2$ (◆) (Reproduced from Ref. [13] with permission).

application owing to the long time to reach the steady state (30 min for hydrogen generation).

3.7
Summary and Outlook

Recent advances in portable powered devices in combination with technological limitations on batteries have sparked interest in hydrogen-based fuel cells. As a result, development of microreactors for hydrogen generation and purification is progressing at a rapid rate within academia and industry. Most work focuses on

Figure 3.4 Photograph of the serpentine patterned silicon wafer with catalyst (Reproduced from Ref. [6] with permission).

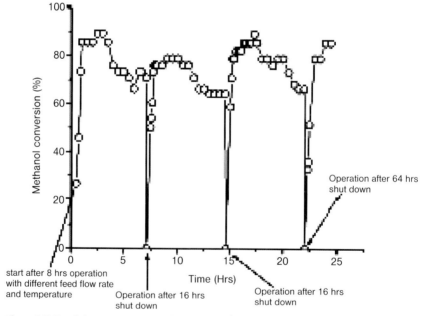

Figure 3.5 Durability test at 260 °C and 0.02 ml min^{-1} with serpentine channeled microreformer (catalyst: Cu/ZnO/Al2O3, catalyst amount: 140 mg) (Reproduced from Ref. [6] with permission).

lower hydrocarbon fuels such as methanol and propane owing to design simplicity and operating conditions. Associated purification of hydrogen from concomitant CO has been investigated extensively using either a water gas shift reactor or palladium membranes. However, issues related to the balance of plant systems integration need to be addressed. Specifically, development of low cost reliable compact blowers and pumps is required for high-pressure operations. Further, addressing fuel cell related issues such as water management will be critical when the system involves closely spaced electronic components. With the continuing advancements however, the prospects of this area appear encouraging.

Acknowledgements

I thank the BP-MIT alliance for funding this project and Prof. Klavs Jensen for his insightful guidance. I also acknowledge Dr. Shashank Dhalewadikar, Dr. Andrea Adamo and Ms. Smeet Deshmukh for their useful suggestions.

References

1 Larminie, J. and Dicks, A. (2000) *Fuel Cell Systems Explained*, 2nd edn, John Wiley & Son, Chichester, p. 189.

2 Holladay, J.D. and Wang, Y. (2004) Review of developments in portable hydrogen production using microreactor technology. *Chemical Reviews*, **104**, 4767–90.

3 Spiegel, C.S. (2007) *Designing and Building Fuel Cells*, 1st edn, Mcgraw-Hill, New York, p. 337.

4 Holladay, J.D., Jones, E.O., Phelps, M. and Hu, J. (2002) Microfuel processor for use in a miniature power supply. *Journal of Power Sources*, **108**, 21–7.

5 Park, H., Malen, J.A., Piggott, W.T., Morse, J.D., Greif, R., Grigoropoulos, C.P., Havstad, M.A. and Upadhye, R. (2006) Methanol steam reformer on a silicon wafer. *Journal of Microelectromechanical Systems*, **15**, 976–85.

6 Kundu, A., Jang, J., Lee, H., Kim, S., Gil, J. and Jung, C. and Oh, Y. (2006) MEMS-Based Micro-fuel processor for application in a cell phone. *Journal of Power Sources*, **162**, 572–8.

7 Kim, T. and Kwon, S. (2006) Design, fabrication and testing of a catalytic microreactor for hydrogen production.

Journal of Micromechanics and Microengineering, **16**, 1760–8.

8 Conant, T., Karim, A. and Datye, A. (2007) Coating of steam reforming catalysts in non-porous multi-channeled microreactors. *Catalysis Today*, **125**, 11–15.

9 Shah, K. and Key, Besser, R. (2007) Issues in the microchemical systems-based methanol fuel processor: energy density, thermal integration, and heat loss mechanisms. *Journal of Power Sources*, **166**, 177–93.

10 Yu, X., Wang, S., Tu, Z. and Qi, Y. (2006) Development of a microchannel reactor concerning steam reforming of Methanol. *Chemical Engineering Journal*, **116**, 123–32.

11 Christian, M. and Kenis, P. (2006) Ceramic microreactors for on-site hydrogen production from high temperature steam reforming of Propane. *Lab on a Chip*, **6**, 1328–37.

12 Tonkovich, A., Yang, B., Perry, S., Fitzgerald, S. and Wang, Y. (2007) From seconds to milliseconds to microseconds through tailored microchannel reactor design of a steam methane reformer. *Catalysis Today*, **120**, 21–9.

13 Wilhite, B.A., Weiss, S.E., Ying, J.Y., Schmidt, M.A. and Jensen, K.F. (2006) High-purity hydrogen generation in a

microfabricated 23 wt% Ag-Pd membrane device integrated with 8:1 $LaNi_{0.95}Co_{0.05}O_3/Al_2O_3$ catalyst. *Advanced Materials*, **18**, 1701–4.

14 Hannemann, S., Grunwaldt, J., Vegten, N., Baiker, A., Boye, P. and Schroer, C. (2007) Distinct spatial changes of the catalyst structure inside a fixed-bed microreactor during the partial oxidation of methane over Rh/Al_2O_3. *Catalysis Today*, **126**, 54–63.

15 Reuse, P., Renken, A., Santo, K., Gorke, O. and Schubert, K. (2004) Hydrogen production for fuel cell application in an autothermal micro-channel reactor. *Chemical Engineering Journal*, **101**, 133–41.

16 Kikas, T., Bardenshteyn, I., Williamson, C., Ejimofor, C., Puri, P. and Fedorov, A. (2003) Hydrogen production in a reverse-flow autothermal catalytic microreactor: from evidence of performance enhancement to innovative reactor design. *Industrial and Engineering Chemistry Research*, **42**, 6273–9.

17 Arana, L., Schaevitz, S., Franz, A., Schmidt, M., Jensen, K. and Microfabricated, A. (2003) Suspended-tube chemical reactor for thermally efficient fuel processing. *Journal of Microelectromechanical Systems*, **12**, 600–12.

18 Deshmukh, S., Mhadeshwar, A. and Vlachos, D. (2004) Microreactor modeling for hydrogen production from ammonia decomposition on ruthenium. *Industrial and Engineering Chemistry Research*, **43**, 2986–99.

19 Ganley, J., Seebauer, E. and Masel, R. (2004) Porous anodic alumina microreactors for production of hydrogen from ammonia. *AIChE Journal*, **50**, 829–34.

20 Srinivas, S., Dhingra, A., Im, H. and Gulari, E. (2004) A scalable silicon microreactor for preferential CO oxidation: performance comparison with a tubular packed-bed microreactor. *Applied Catalysis A: General*, **274**, 285–93.

21 Ko, E., Park, E., Seo, K., Lee, H., Lee, D. and Kim, S. (2006) $Pt-Ni/\gamma-Al_2O_3$ catalyst for the preferential CO oxidation in the hydrogen stream. *Catalysis Letters*, **110**, 275–9.

22 Uysal, G., Akin, A., Onsan, I. and Yildirim, R. (2006) Hydrogen clean-up by preferential CO oxidation over Pt-Co-Ce/MgO. *Catalysis Letters*, **108**, 193–6.

23 Wilhite, B., Schmidt, M. and Jensen, K. (2004) Palladium-based micromembranes for hydrogen separation: device performance and chemical stability. *Industrial and Engineering Chemistry Research*, **43**, 7083–91.

24 Keurentjes, J., Gielens, F., Tong, H., Rijn, C. and Vorstman, M (2004) High-flux palladium membranes based on microsystem technology. *Industrial and Engineering Chemistry Research*, **43**, 4768–72.

25 Zhang, Y., Gwak, J., Murakoshi, Y., Ikehara, T., Maeda, R. and Nishimura, C. (2006) Hydrogen permeation characteristics of thin palladium membrane prepared by microfabrication technology. *Journal of Membrane Science*, **277**, 203–9.

26 Ye, S., Tanaka, S. and Esashi, M. (2005) Thin palladium membrane microreactors with oxidized porous silicon support and their application. *Journal of Micromechanics and Microengineering*, **15**, 2011–18.

27 Bogdanovic, B., Brand, R., Marjanovic, A., Schwickardi, M. and Tolle, J. (2000) Metal-doped sodium aluminum hydrides as potential new hydrogen storage materials. *Journal of Alloys and Compounds*, **302**, 36–58.

28 Richardson, B., Birdwell, J., Pin, F., Jansen, J. and Lind, R. (2005) Sodium borohydride based hybrid power system. *Journal of Power Sources*, **145**, 21–9.

4
Micro Fuel Cells

Joshua L. Hertz and Harry L. Tuller

4.1
Introduction

As detailed in earlier chapters, there are a number of benefits associated with chemical micropower systems. Fuel cells, first discovered by Grove over 160 years ago [1], are in some ways the simplest means by which to obtain usable chemical micropower since they directly convert chemical to electrical energy. Accordingly, a large number of groups in academia and industry have begun investigating means to produce portable micro-fuel cells. Cowey *et al.* estimate [2] that over 1700 programs worldwide are demonstrating sub-1 kW fuel cells (with varying degrees of portability). Polymer electrolyte membrane fuel cells have been the focus of the vast majority of these efforts given that they operate at relatively low temperatures. Some of these appear to be nearing commercialization [3]. However, because this type of fuel cell is limited in its operation to hydrogen fuel, higher temperature alternatives, including the more fuel flexible solid oxide fuel cell (SOFC), are also receiving attention. However, no fuel cell system, large or small, has yet been widely adopted, primarily due to high capital cost [4].

The main benefit of fuel cells for chemical micropower generation is the simplicity of energy transduction. The direct conversion to electrical energy promises high efficiency and reliability. There are no moving parts or thermal conversion routes required by the technology. In addition, the fuel flexibility of SOFCs potentially allows direct use of hydrocarbons, alcohols, and other more complex and/or less pure options. On the other hand, significant engineering hurdles remain in producing micro-fuel-cell power sources. Many of these stem from the requirement that the fuel cell forms a stable membrane, sealing off a fuel chamber from an air chamber (an exception to this is described in Section 4.5.1). Challenges also remain in ensuring the long term chemical, mechanical and thermal stability of the materials.

In this chapter, the basic operating principles of fuel cells and the mechanisms that limit performance are first presented. Included in this discussion are the challenges posed by miniaturizing such devices. This is followed by a discussion

Microfabricated Power Generation Devices. Edited by Alexander Mitsos and Paul I. Barton
Copyright © 2009 WILEY-VCH Verlag GmbH & Co. KGaA, Weinheim
ISBN: 978-3-527-32081-3

of the different types of fuel cell technologies that are most likely to be deployed for micropower, that is, the solid oxide and the polymer electrolyte fuel cell. For each of these technologies, the challenges and opportunities posed by transforming these devices into portable form will be discussed alongside current research. Next, several emerging types of fuel cells well suited for miniaturization are described. Finally, we conclude with an outlook to the future of fuel cells for micropower.

4.2
Principles of Operation

4.2.1
Electrochemical Potential

At heart, the fuel cell is not very different from a battery in that both directly transform chemical into electrical energy via a galvanic cell. As in any electrochemical cell, the presence of an electronically insulating, ion conducting electrolyte creates a coupling between the chemical potential and the electrical potential. The main difference between fuel cells and batteries is that fuel cells use continually replenished reactants instead of finite reactants contained within the electrodes. The benefit of the fuel cell arrangement is that the energy storage function of the device is decoupled from the electrode function, providing more liberty to choose the reductant and oxidant, optimizing for increased energy density, life and/or reduced cost. Though, in principle, the fuel cell reactants could be liquid or even solid, they are nearly always gaseous (notable exceptions will be described later in this chapter).

The operation of a fuel cell is initiated by flowing a reductant stream (i.e. fuel) and an oxidant stream across opposite sides of the electrolyte. This creates a chemical potential gradient for the mobile ion, normally a proton or oxygen ion, which must be a constituent of the reductant or oxidant (or both). Equilibrium is reached when the electrochemical potential, $\tilde{\mu}_i$, of mobile ion i,

$$\tilde{\mu}_i = \mu_i + n_i \cdot F \cdot \phi \tag{4.1}$$

is uniform across the electrolyte, where μ_i is the chemical potential, n_i is the charge number of the species, F is Faraday's constant $= 9.6484 \times 10^4 \, C \, mol^{-1}$, and ϕ is the electrical potential. This is achieved by the chemical potential gradient being matched by an electrical potential gradient, equivalent in magnitude, but opposite in sign. This electrical potential difference can then be used to drive an electronic current through an external load, powering the device.

The electrical potential difference, or voltage, across an electrochemical cell can be derived by equating the Gibbs free energy due to the chemical potential gradient

$$\Delta G = \Delta G^0 + RT \ln\left(\frac{a_{red}}{a_{ox}}\right) \tag{4.2}$$

with the Gibbs free energy due to an electrical potential gradient

$$\Delta G = -n \cdot F \cdot \Delta\phi \tag{4.3}$$

to derive the Nernst equation:

$$\Delta\phi = \Delta\phi^0 - \frac{RT}{n \cdot F} \ln\left(\frac{a_{red}}{a_{ox}}\right) \tag{4.4}$$

where ΔG is the Gibbs free energy; ΔG^0 is the standard molar reaction free energy; R is the universal gas constant = 8.3144 J mol^{-1} K^{-1}; T is the absolute temperature; a_{red} and a_{ox} are the chemical activities of the species at the reductant/electrolyte (anode) and oxidant/electrolyte (cathode) interfaces, respectively; and $\Delta\phi$ is the electrical potential difference across the cell. $\Delta\phi^0$, related to ΔG^0 by Equation 4.3, is the potential difference when the chemical species on either side of the electrolyte are in their standard states, for then the chemical activities are defined as 1 and $\Delta\phi = \Delta\phi^0$. It therefore represents the cell potential at equilibrium – the zero-current or open circuit voltage – at standard temperature and pressure.

To derive usable power, some of the electromotive force must be used to drive a current through an external load. Whenever current flows, the system is taken out of equilibrium and the actual output voltage is reduced. The sources of the reduced voltage, called overpotentials or polarizations and given the symbol η, are commonly divided into three categories: ohmic, concentration, and activation polarization and are discussed in the following.

4.2.2
Ohmic Polarization

The ohmic polarization, η_{ohm}, is the voltage drop that occurs due to the resistance of the electrolyte to the flow of ions. An ohmic resistance, R, implies a linear relationship between voltage and current, i, and is given by:

$$\eta_{ohm} = i \cdot R \tag{4.5}$$

The resistance is analogous to, for example, electronic resistance in metals with equivalently defined conductivity, σ_{ion}, and dependence on the geometric parameters of path length, l, and cross-sectional area, A:

$$R = \frac{l}{\sigma_{ion} \cdot A} \tag{4.6}$$

A great deal of research is focused on increasing electrolyte conductivity. The need and means by which to improve conductivity is dependent upon the particular fuel cell technology and will be discussed in later sections. Reducing the path length of the ionic current also reduces the resistance. Thin film electrolytes, as would likely be used within a micro fuel cell, thus offer promise for reducing ohmic polarization.

4.2.3
Activation Polarization

The activation polarization arises from thermodynamic irreversibilities in the electrochemical reactions. The electrochemical half-reactions at the electrodes can be written generically as

$$Ox + n \cdot e^- \Leftrightarrow Re \tag{4.7}$$

where Ox is the oxidized species, Re is the reduced species, n is the number of electrons transferred, and the direction of the reaction is forward at the cathode and reverse at the anode. One of the species, Ox or Re, must be in the gas phase while the other is the mobile ion in the electrolyte. Because the mobile ion comes from a substance external to the electrode, the overall half-reaction necessarily includes a number of sub-reactions. These include the gas phase diffusion of the reactant to the electrode or electrolyte surface; adsorption; dissociation; ionization; surface diffusion to the electrochemically active site; incorporation of the ion into the electrolyte and, on either the anode or cathode, reaction of the ion to create the waste product. This complex series of steps, illustrated for an oxygen reactant at a platinum cathode/zirconia oxygen ion conducting electrolyte interface in Figure 4.1, is made more complex by the fact that some of the steps occur in parallel or may be arranged in a different order. Poor diffusion in the gas phase is generally treated as concentration polarization, and is discussed in greater detail below. A rate-limiting reaction step in any of the other subreactions is termed activation polarization.

Because of the complexity of this series of reactions, physical modeling has remained a difficult task. To further complicate matters, the electrode kinetics is highly dependent upon operating conditions, materials composition, and surface structure and chemistry. As a result, the identity and speed of the rate-limiting step(s) often vary among laboratories and even from experiment to experiment. Empirically, the activation polarization is found to follow Tafel kinetics

$$\phi = a + b \log i \tag{4.8}$$

with voltage, ϕ, depending logarithmically on current, i, and experimentally derived fitting parameters a and b. This can be explained using the Butler–Volmer model of interfacial kinetics.

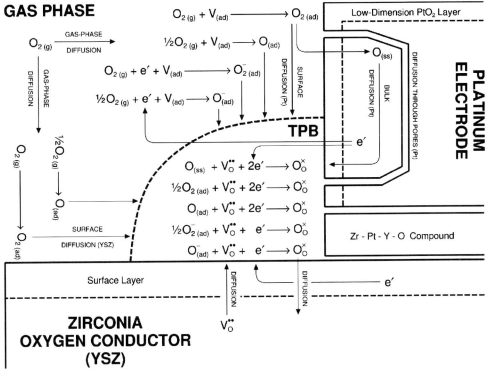

GAS PHASE

$O_{2\,(g)} + V_{(ad)} \longrightarrow O_{2\,(ad)}$

Low-Dimension PtO$_2$ Layer

GAS-PHASE
$O_{2\,(g)} \xrightarrow{\text{DIFFUSION}}$ $\tfrac{1}{2}O_{2\,(g)} + V_{(ad)} \longrightarrow O_{(ad)}$

$O_{(ss)}$

$O_{2\,(g)} + e' + V_{(ad)} \longrightarrow O_{2\,(ad)}^-$

$\tfrac{1}{2}O_{2\,(g)} + e' + V_{(ad)} \longrightarrow O_{(ad)}^-$

TPB

e'

$O_{(ss)} + V_O^{\bullet\bullet} + 2e' \longrightarrow O_O^x$

$\tfrac{1}{2}O_{2\,(ad)} + V_O^{\bullet\bullet} + 2e' \longrightarrow O_O^x$

$O_{(ad)} + V_O^{\bullet\bullet} + 2e' \longrightarrow O_O^x$

$\tfrac{1}{2}O_{2\,(ad)}^- + V_O^{\bullet\bullet} + e' \longrightarrow O_O^x$

Zr - Pt - Y - O Compound

$O_{(ad)}^- + V_O^{\bullet\bullet} + e' \longrightarrow O_O^x$

SURFACE
DIFFUSION (YSZ)

Surface Layer

$V_O^{\bullet\bullet}$

ZIRCONIA OXYGEN CONDUCTOR (YSZ)

PLATINUM ELECTRODE

Figure 4.1 The many reaction mechanisms, occurring in series and in parallel, that bring an oxygen reactant into the surface of a zirconia solid oxide fuel cell electrolyte in the presence of a platinum electrode. Reprinted from Ref. [5].

A model for this can be derived from kinetic rate theory [6], assuming just one of the subreactions – generally the one with the largest activation barrier – is rate limiting. From this derivation, the Butler–Volmer equation results:

$$i = i_0 \left[\exp\left(-\frac{\alpha F \eta_{\text{act}}}{RT}\right) - \exp\left(\frac{(1-\alpha) F \eta_{\text{act}}}{RT}\right) \right] \tag{4.9}$$

where α is the transfer coefficient (≈ 0.5), a unit-less parameter that corrects for asymmetry in the shape of the energy barrier; i_0 is the exchange current; and η_{act} is the activation polarization. From this equation, it can be seen that i_0 essentially represents the current that flows, equally in both directions, at equilibrium. The first exponential describes the forward current and the second exponential describes the backward current. At large overpotentials, one of the two terms in Equation 4.9 will become insignificant and the resultant current–potential relationship will follow that described by Equation 4.8, the empirically derived Tafel equation. At low overpotentials, the Butler–Volmer equation can be linearized (since $\exp(x) \approx$

$1 + x$ for small x) such that a charge transfer resistance, $R_{CT} = i/\eta_{act}$ can be derived.

Reducing activation polarization is accomplished in two ways. First, the materials chosen for the electrode and/or electrolyte can be more catalytically active. This reduces the energy of the activated state of the rate-limiting step, increasing i_0 and thereby the kinetic rate. Second, the concentration of sites that are active to the electrochemical reaction can be increased. Since the half-reactions in a fuel cell involve a gaseous species, an electron, and an ionic species, the reactive sites must be in some local vicinity to the gas phase, the electrode phase, and the electrolyte phase [7]. This fact was discovered as early as Grove's original paper [1] and has since been validated by a number other studies [7]. This region of contact is known as the triple phase boundary (TPB), shown as the dash-enclosed region of Figure 4.1. Increasing the TPB length per unit area is accomplished by using composite electrodes, with interpenetrating networks of electrolyte, electrode, and pore phases (see Figure 4.2). These microstructures can be further optimized for composition and grain/pore size.

As the common intersection of three volumes, the TPB is ostensibly a one-dimensional region. It nevertheless has some volume, due to the non-zero electron conductivity of the electrolyte, ionic conductivity of the electrode, and gas permeability of either solid phase. Any of these will broaden, to some extent, the region over which reactions can occur, giving some width and depth to the TPB. This idea can be carried further by intentionally using a material that is highly conductive to both ions and electrons as the electrode. Such mixed ionic electronic conductors (MIECs) relieve the need for a TPB and enable the half-reactions to occur everywhere along the gas–MIEC surface.

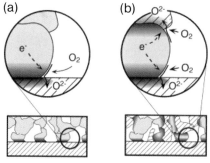

Figure 4.2 Electrochemically active sites (shown here for a solid oxide fuel cell cathode, but relevant to all fuel cell types) are only found where the white gas phase, gray electrode phase, and hashed electrolyte phase meet. (a) The single phase electrode has active sites only at the planar interface. (b) The composite electrode has active sites throughout the thickness, assuming interconnectedness of the phases. The composite is expected to have reduced activation polarization. Reprinted from Ref. [7].

4.2.4
Concentration Polarization

Concentration polarization is due to insufficient gas phase diffusion of reactants. When the electrochemical reactions proceed at a sufficiently high rate, reactants become depleted near the electrode/electrolyte interfaces. By Raoult's law, chemical activity is proportional to the concentration. The Nernst equation (Equation 4.4) therefore demands that reduced concentration leads to reduced voltage. Writing η_{conc} for the electrical potential drop due to the concentration gradient,

$$\eta_{conc} = \frac{RT}{nF} \ln\left(\frac{a^{int}}{a^{react}}\right) \tag{4.10}$$

where a^{int} is the chemical activity at the electrode/electrolyte interface and a^{react} is the chemical activity at a far distance from the interface, equal to that in the original reactant stream. The dependence of this polarization on current can be estimated by modeling the flux of the chemical species in the gas phase as Fickian diffusion through a boundary layer of thickness δ. The current density flowing through the cell, j, equals the flux of species i per unit cross-sectional area multiplied by the charge per mole of the species:

$$j = -nFD\nabla\mu_i \approx -nFD\frac{a^{react} - a^{int}}{\delta} \tag{4.11}$$

where D is the diffusivity and $\nabla\mu$ is the chemical potential gradient. The maximum flux through the boundary layer will occur when the reactant is completely depleted at the electrolyte interface, for then the largest gradient in activity is found. Defining a limiting current density, j_L, for this situation gives

$$j_L = -nFD\frac{a^{react}}{\delta} \tag{4.12}$$

Rearranging Equations 4.11 and 4.12 and inserting them into Equation 4.10 yields an expression for the concentration polarization as a function of the current flowing through the cell:

$$\eta_{conc} = \frac{RT}{nF} \ln\left(1 - \frac{j}{j_L}\right) = \frac{RT}{nF} \ln\left(1 - \frac{i}{i_L}\right) \tag{4.13}$$

where the second equality uses the current instead of current density since the ratios are equivalent.

In order to reduce the concentration polarization, there must be rapid diffusion of the gas reactants to the electrode/electrolyte interfaces. This implies a need for electrodes that have high open porosity so that gases can quickly reach the triple phase boundaries. A well-designed fuel cell has sufficient porosity so that concentration polarization becomes a significant loss mechanism only at high currents.

Special attention must be paid to the concentration polarization at the electrode to which the mobile ion travels because the reaction products are created there. Diffusion of the reactants to the electrolyte interface necessarily implies diffusion of the products away from that interface. This is particularly important when the waste stream is not gaseous. Low-temperature fuel cells can produce liquid water wastes that must be quickly distributed away from the electrodes to prevent "flooding" [8]. This can be done by treating the electrode surfaces to make them hydrophobic.

4.2.5
Practical Device Performance

Devices can suffer two loss mechanisms that concern leakage mechanisms through the electrolyte. The first is the crossover of the reactants from one side of the electrolyte to the other, either through pores or by some other permeation mechanism. This reduces the chemical potential gradient across the electrolyte, lowering the voltage output from the cell. It also wastes reactant so that the overall efficiency of the system is decreased. The second relates to leakage by electron (or hole) conduction through the electrolyte. Because no material is perfectly insulating to electron conduction, some electron current travels through the electrolyte rather than the external circuit. The current available to the load is thereby reduced. A generalization of the Nernst equation can be used to account for electron conduction through the electrolyte [9]:

$$\Delta\phi = \Delta\phi^0 - \frac{RT}{n \cdot F} \int_{a_{ox}}^{a_{red}} \frac{\sigma_{ion}}{\sigma_{ion} + \sigma_{electron}} d\ln a \tag{4.14}$$

where σ_{ion} and $\sigma_{electron}$ are the ionic and electronic conductivities, respectively, of the electrolyte. These two loss mechanisms are active even when no current flows through the external load, and so can be perceived by a less-than-theoretical voltage when the system is held at open circuit.

Additional losses found within fuel cell systems come from inefficiencies in the balance of plant components that supply the reactants, collect the current, condition the electrical output to desired (alternating) voltage and/or current, perform thermal management, and preprocess the fuel into usable form. These are not considered further in this chapter.

A common way of displaying the performance of a fuel cell is by plotting the voltage against the current, as in Figure 4.3. When no current flows through the external circuit, then the potential is the open circuit voltage, V_{OC}. Theoretically, this is the same as the Nernst voltage for any given oxidant/fuel combination. In practice, V_{OC} is reduced by the leakage mechanisms described above. If current is allowed to flow, the various polarization mechanisms will lead to a voltage that decreases monotonically with increasing current. At the extreme, when there is zero resistance in the external circuit, there can be no voltage drop and the resultant current is defined as the short circuit current, I_{SC}.

At any point on the performance curve, the voltage output is equal to the ideal, Nernst potential minus the total loss, equal to the sum of all of the overpotentials.

$$\phi_{operating} = \phi_{Nernst} - \eta_{tot} = \phi_{Nernst} - (\eta_{ohm} + \eta_{act} + \eta_{conc}) \qquad (4.15)$$

In general, only one of the polarization mechanisms is dominant at any specified operation point, with activation polarization likely to be dominant at low, ohmic polarization at intermediate and concentration polarization at high currents. In a plot like Figure 4.3, the dominant loss mechanism regions can often be identified by convex, linear, and concave curvature of the activation, ohmic, and concentration polarization regions, respectively.

The performance of a fuel cell can be improved by increasing its cross-sectional area. For comparison purposes, materials and device parameters are normalized to this area. The charge transfer resistance and the resistance of the electrolyte are often given as area specific resistance (ASR) with units of $\Omega\,cm^2$. Similarly, currents and powers are often given as current and power (area) densities, as on the x-axis of Figure 4.3.

The power of any electric device is the product of voltage and current, $P = V \cdot I$. While the voltage decreases monotonically as current increases, the power output goes through a roughly parabolic curve, with some operation point that gives maximal power output, P_{max}. Device performance curves are typically presented as overlaid plots of voltage and power density versus current density.

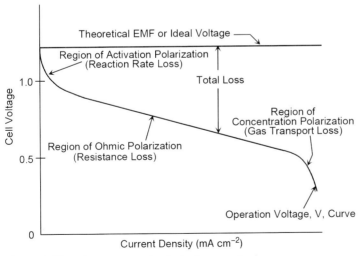

Figure 4.3 The voltage vs. current output of a generalized fuel cell with typical regions for activation, ohmic and concentration polarization indicated. Reprinted from Ref. [8].

The operation point of maximum power output is not necessarily the most desirable, because it is generally not the operating point of maximum efficiency. One may choose, depending on the application, to operate a fuel cell at the point of maximum power output, the point of maximum efficiency, or between the two at a point of optimized operation cost [2]. These operational optimization issues will be discussed in more detail in later chapters. The maximum possible efficiency of a fuel cell comes from the ratio of the maximum possible useful energy output to the stored chemical energy input

$$\text{efficiency} = \frac{\Delta G}{\Delta H} \qquad (4.16)$$

This efficiency can only be strictly met when a fuel cell operates reversibly. The overall irreversibility of the device can be found in the ratio of the operating voltage to the ideal voltage so that the efficiency of the fuel cell can be expressed as [8]

$$\text{efficiency} = \frac{\phi_{\text{operating}} \cdot \Delta G}{\phi_{\text{Nernst}} \cdot \Delta H} \cdot \delta \qquad (4.17)$$

where the additional factor δ is the proportion of the fuel that is electrochemically oxidized as opposed to that which is chemically oxidized by gas leakage or is sent unused to the exhaust. Real systems may, of course, use combined cycle generators or other methods to use the fuel energy more efficiently.

It follows from Equation 4.17 that operating the fuel cell at lower current density, where the operating voltage is closer to the Nernst potential, increases the efficiency. On the other hand, low current density implies low power density and so a fuel cell with larger cross-sectional area is required to create the same total power. The key to creating fuel cells with both high efficiency and large power density is therefore to reduce the polarizations so that at a sufficient power output level, the actual voltage is as close as possible to the ideal voltage.

For nearly all technologically important fuel cells, air is the oxidant because it can be used without added cost or system weight. Hydrogen is often used as the reductant (i.e. fuel) for a number of reasons. First, it is highly reactive and so places fewer burdens on the catalytic properties of the anode material. Second, there are electrolytes with high conductivity for hydrogen ions. Thus, for fuel cells employing these electrolytes, pure hydrogen is the simplest fuel and does not require further fuel processing. Third, it reacts only to form water, without the possible complications of solid decomposition products such as carbon. On the other hand, hydrogen is expensive, difficult to store in a dense manner, and prone to leakage. For these reasons, it is beneficial to feed hydrocarbon fuels directly to the fuel cell. This is possible if the fuel cell operates in a regime where the carbon–carbon and carbon–hydrogen bonds can be broken and the fuel oxidized.

With hydrogen as the fuel and oxygen or air as the oxidant, the overall chemical reaction that powers the fuel cell is

$$2H_2 + O_2 \Rightarrow 2H_2O \qquad (4.18)$$

The electrolyte splits this reaction into two half-reactions, the details of which depend on which mobile ion is being transferred through the electrolyte (see Figure 4.4). Nevertheless, reduction always occurs at the cathode and oxidation at the anode and one of the ionic species then migrates through the electronically insulating electrolyte. If the electrolyte transports hydrogen ions, such as in polymer electrolyte membrane fuel cells and phosphoric acid fuel cells, then the half-reactions at the anode and cathode are, respectively,

$$
\begin{aligned}
& H_2^{(g)} \Rightarrow 2H^+ + 2e^- \\
\text{and} \quad & O_2^{(g)} + 4H^+ + 4e^- \Rightarrow 2H_2O
\end{aligned}
\tag{4.19}
$$

These can be compared with the generic half-reaction of Equation 4.7. To connect the half-reactions, the hydrogen ion passes through the electrolyte and the electron through the external load. If the electrolyte transports oxygen ions, as in solid oxide, molten carbonate, and alkaline fuel cells, then the half-reactions at the anode and cathode are, respectively,

$$
\begin{aligned}
& H_2^{(g)} + O^{2-} \Rightarrow H_2O + 2e^- \\
\text{and} \quad & O_2^{(g)} + 4e^- \Rightarrow 2O^{2-}
\end{aligned}
\tag{4.20}
$$

and the oxygen ion (perhaps as part of a complex ion) passes through the electrolyte to connect the half-reactions. After the fuel and oxygen react, the waste

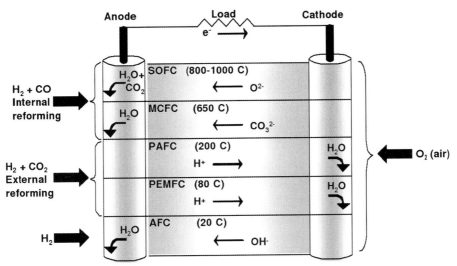

Figure 4.4 The common fuel cell types, showing the motions of the mobile ions through the electrolyte and the associated fuel types. AFC = alkaline fuel cell, PEMFC = polymer electrolyte membrane fuel cell, PAFC = phosphoric acid fuel cell, MCFC = molten carbonate fuel cell, and SOFC = solid oxide fuel cell. Adapted from Ref. [10].

products are emitted as exhaust. These processes are illustrated for the various fuel cell technologies in Figure 4.4.

4.2.6
Micro Fuel Cells

There have been efforts, for reasons explained below, to use thin film electrolytes in traditional, large scale fuel cells [11, 12]. These are sometimes referred to as thin film fuel cells; however, they are not of a suitable size or design for a battery replacement and will not be discussed here.

As mentioned in Section 4.2.2, one of the main sources of loss in fuel cells stems from insufficient ionic conductivity in the electrolyte. Equation 4.6 indicates that the ohmic polarization is reduced by a reduction in the path length of the ionic charge through the electrolyte. Thus, thin film electrolytes have the potential to reduce ohmic polarization. This point is illustrated for solid oxide fuel cells in Figure 4.5, which shows an Arrhenius plot of the ionic conductivity for several well-known oxygen ion conductors. Also shown is a typical goal value for the area specific resistance of the ohmic polarization, $R_0 = 0.15\,\Omega\,cm^2$. With a specified ASR, the conductivity is directly correlated with a required thickness according to Equation 4.6. Therefore, on the right axis is given the maximum electrolyte thickness that provides an ASR at or below the goal. Electrolytes produced by traditional ceramic means of powder processing typically result in thicknesses of at least hundreds of μm. From the figure, it can be seen that this requires operation temperatures of over 800 °C for $(ZrO_2)_{0.9}(Y_2O_3)_{0.1}$ and upwards of 600–700 °C for the other materials. Using thin film processing methods to produce the electrolyte,

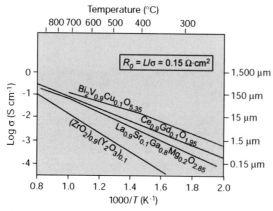

Figure 4.5 The ionic conductivity of various oxide ion conductors is plotted versus inverse temperature. With a goal area specific resistance of $0.15\,\Omega\,cm^{-2}$, a one-to-one correlation between conductivity and thickness can be made. Adapted from Ref. [10].

on the other hand, can result in dense ceramics with thicknesses of only hundreds of nm or possibly less. This potentially allows the fuel cell to operate at 300 °C or less for the higher conductivity materials and even for the popular, but less conducting zirconia material, operation at 400 °C or less may be possible.

The previous discussion assumed uniform current density within the electrolyte. Since charge transfer occurs only near the triple phase boundaries, the current in the near-surface region of the electrolyte is constricted into these regions. The triple phase boundaries are found at the edges of electrode particles, so whenever the electrolyte thickness approaches the electrode particle size, the added constriction resistance becomes significant. The resistance then ceases to decrease linearly as electrolyte thickness is reduced [13], however the ohmic losses are still likely to be insignificant when compared to the activation polarization losses [14]. As a case in point, Xu *et al.*, reported a SOFC using a 10 µm thick yttria-stabilized zirconia (YSZ) electrolyte [15]. At temperatures between 600 and 750 °C, the cell resistance was only about 20% due to the electrolyte resistance, with the remaining 80% due to electrode polarization.

For activation polarization, there is a significant geometric disadvantage to using thin film electrodes. The relevant geometric parameter for activation polarization is the triple phase boundary length per unit of projected surface area. Traditionally processed electrodes are thick and porous, and thereby have a large TPB length as well as high catalyst surface area that is accessible through the thickness. Figure 4.6 illustrates this point with a typical SOFC anode and cathode. The figure shows that the resistances of the electrodes decrease as the thickness is increased, not reaching optimal values until the electrodes are greater than 10 µm thick. Thin films are typically not porous, thick, or composites. Therefore, it can be difficult to produce a thin film electrode with low activation polarization. Generally, the cathode and anode reactions are temperature activated so, all else being equal, thin

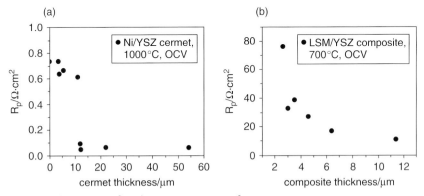

Figure 4.6 The area specific polarization resistance of (a) a composite anode and (b) a composite cathode as a function of the layer thickness. The polarization increases significantly at thickness less than 10 µm. Reprinted from Ref. [16].

film electrodes would need to be operated at *increased* temperature compared to their traditional counterparts. Therefore, the activation polarization in micro fuel cells is likely to be the biggest electrochemical hindrance to achieving reasonable power output.

The third polarization mechanism, concentration polarization, is not likely to be an issue for micro fuel cells [14]. Thin film electrodes decrease the diffusion distance for the reactants in the gas phase and so micro-fuel cells are likely to have low concentration polarization.

One of the main difficulties in fabricating a fuel cell from thin films concerns the need for reactants to access the triple phase boundaries of the anode and cathode. Using the planar architecture typical of thin film devices requires the use of a substrate that is either innately porous or made so by etching. On the other hand, the electrolyte must be dense in order to prevent leakage of the gases from one chamber to the other. Therefore, the fuel cell stack must form a mechanically robust, freestanding membrane across pores or other holes in the electrodes. This may be particularly troublesome during thermal cycling, since the coefficient of thermal expansion of traditional substrates like silicon ($\alpha = 3 \times 10^{-6}$ K^{-1}) is substantially different from that of traditional fuel cell materials, like YSZ ($\alpha = 10 \times 10^{-6}$ K^{-1}) [17] or Nafion ($\alpha = 1.5 \times 10^{-4}$ K^{-1}) [18]. Nevertheless, design diagrams to identify reliable geometries have been made [17–19], using assumptions of the mechanical properties of the films and the fracture mechanics of the structures. In addition to thermal cycling, mechanical strain occurs due to chemical effects. Considerable chemical expansion occurs with uptake or loss of oxygen ions in SOFC materials and of water in PEM materials.

4.3
Solid Oxide Fuel Cells

4.3.1
Basic Principles

A fuel cell using a ceramic electrolyte was first reported by Baur and Preis in 1937 [20]. This was the first of a class of fuel cells now termed solid oxide fuel cells, characterized by a solid metal oxide electrolyte that is conductive to oxygen ions. The use of a ceramic for the electrolyte is advantageous in that these materials are generally highly chemically and thermally stable. SOFCs are, in some sense, the only fuel cell technology with all of the components made of solid materials. MCFCs, PAFCs, and AFCs have liquid electrolytes and the PEMFCs have a hydrated polymer electrolyte so that water management is still a key concern (acronyms defined in the caption to Figure 4.4). The use of solid components gives the SOFC structural stability and eliminates the corrosion and containment issues associated with liquid electrolytes.

The drawback to a solid electrolyte is that there are relatively few solid materials that have ionic conductivities on a par with liquid electrolytes. The ceramic oxygen conductors all have highly thermally activated conductivities, with values only approaching that of liquid electrolytes at very high temperatures (>600 °C). The necessity of operating at such elevated temperatures drives the engineering of the other SOFC components. In particular, the device housing and the sealant between the fuel and oxidant chambers must be stable for extended times at operating temperature and during power up/down thermal cycling. The additional expenses associated with high-temperature platforms have been one of the main contributors to high SOFC capital costs. Research into ways to reduce the operating temperature is highly active [21].

There are benefits to operating at elevated temperatures, however. First, the electrochemical reaction kinetics are accelerated, thus reducing the catalytic activity requirements of the electrode materials. Whereas lower-temperature fuel cells need highly catalytic platinum, SOFCs can use less costly substitutes [22]. Second, at high temperatures it becomes easier to break carbon–carbon bonds and thus hydrocarbons can be directly utilized at the anode [23]. Third, the high temperatures of the exhaust gases can be used to increase efficiency, using a combined cycle or cogeneration. One of the goals of current SOFC research is to engineer a device that operates at a sufficiently high temperature to use cheap catalysts and hydrocarbon fuels, yet at a low enough temperature to use lower cost sealant and housing components.

The primary requirement of the electrolyte is to be highly conductive to oxygen ions and highly resistive to electrons. Electrolytes in an SOFC are exposed to both the oxidizing cathode and reducing anode environment and must therefore be stable at high temperatures in both environments. The most popular electrolyte is stabilized zirconia (zirconium oxide) given its excellent chemical and electrochemical stability. The main limitation of zirconia is its somewhat low ionic conductivity at lower temperatures. Research into electrolytes with improved conductivity has been active for many years and while some have been identified, for example doped ceria and lanthanum gallate, they tend to exhibit lower chemical or electrochemical stability [24].

The anode must satisfy the requirements of high electron conductivity and stability in a reducing environment. Ideally, it should also be catalytic towards fuel oxidation, have a similar thermal expansion coefficient to the electrolyte and be inexpensive. Base metals satisfy these requirements and nickel is the traditional choice [22]. Typically, a porous composite called a cermet is made of nickel and the electrolyte for three reasons: to increase the TPB length, to reduce high-temperature agglomeration of the nickel particles, and to reduce thermal expansion mismatch.

The cathode has the same requirements as the anode material, except that it works in an oxidizing environment and is ideally catalytic towards the reduction reaction. Base metals will quickly oxidize in this environment so either a noble metal or metal oxide conductor must be used. The latter is typically used for both morphological stability and cost reasons, traditionally lanthanum–strontium manganite ($La_{1-x}Sr_x$)MnO_3 (LSM) [22]. Like the anode material, it is often made into a composite with the electrolyte. Variations in composition, particularly with respect to replacement of the Mn with other transition metals or alkaline earth metals, are being investigated to improve the kinetic reaction rates [21].

The electrolyte in an SOFC conducts species from the oxidant instead of the fuel, so potentially any fuel that combines with oxygen can be used. The disadvantage of this arrangement is that the exhaust gases are released into the fuel stream instead of the air. This inherently limits the fuel utilization, since increased fuel utilization implies increased concentration polarization [25]. This is in contrast to fuel cell technologies where exhaust is produced on the air/cathode side. Airflow can then be increased to purge the exhaust since air is cheap, both literally and in the sense of zero added system volume and weight.

4.3.2
Micro Solid Oxide Fuel Cells

In addition to the issues discussed in Section 4.2.6, the main obstacle for micro solid oxide fuel cells (μSOFCs) lies in maintaining an efficient energy balance despite heat losses to the substrate and packaging. While there is no inherent size dependence of efficiency for a fuel cell, there are such dependences on the heat lost to the surroundings. A portable device implies that the user-contactable packaging must be at or near room temperature. The high-temperature, active portion of a μSOFC is in close proximity to the packaging, so significant heat losses may occur. Heat loss by conduction to the substrate is expected to dominate for simple geometries, with the loss perhaps exceeding the electrical power produced [17, 19]. Efficient thermal insulation schemes, including advanced vacuum packaging, may be required to improve the efficiency of the devices [26]. In addition, cycling the device between ambient and operating temperatures places a heavy burden on the thermomechanical stability of the anode/electrolyte/cathode membrane. A few research efforts have detailed methods to produce large area, thin film membranes of materials typical to SOFCs [27–29].

Lawrence Livermore National Laboratory (LLNL) has had a concerted research effort into micro fuel cells integrated onto silicon substrates, with a similar design used to produce polymer electrolyte and solid oxide fuel cells [30–33]. The general design is shown in Figure 4.7a. A start-up company, UltraCell Corporation (Livermore, CA), appears to be licensing the technology [34].

The design employs deep reactive ion etching to create "pores" in the silicon substrate. A low stress silicon nitride membrane covers this frame and provides a dense substrate on which to deposit the thin films. After the fuel cell films are deposited, pores are etched into the silicon nitride membrane, as seen in Figure 4.7b. The membrane size is given as $2 \times 2\,mm^2$, though it is unclear whether this is the area of the etched silicon substrate or the total area exposed through the silicon nitride pores.

The design includes an integrated electrical resistance heater. This is used to turn on the fuel cell by heating the device to operating temperatures and is powered by a rechargeable micro-battery packaged with the device. The authors state that once the fuel cell is operating, the resistive heater will be powered by

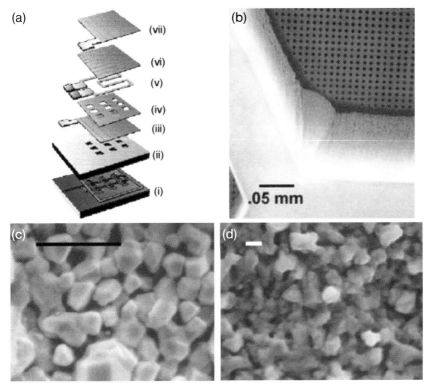

Figure 4.7 Different views of the LLNL µSOFC. (a) A schematic view with (i) micromachined substrate manifold, (ii) thermal and electrical insulation, (iii) thin film anode, (iv) electrical insulating layer, (v) electrical resistance heater, (vi) thin film electrolyte, and (vii) thin film cathode. (b) Viewing from the bottom, the silicon manifold (light gray) is formed into a hexagonal support structure, supporting the silicon nitride (dark gray) insulation layer with smaller pore structure. (c) Top view of a porous, sputtered nickel anode (bar = 1 µm). (d) Top view of a porous, sputtered silver cathode (bar = 1 µm). Reprinted from Ref. [31].

the fuel cell itself [33], though presumably improved thermal insulation would allow steady state operation without wasting energy through a resistive heater.

The µSOFC uses a porous nickel film for the anode (Figure 4.7c) and a porous silver film (Figure 4.7d) for the cathode. Porosity was achieved in the electrode films by sputtering them at high pressures and at elevated substrate temperature [35]. The electrolyte is a dense thin film of YSZ, formed by sputtering from a $(ZrO_2)_{0.94}(Y_2O_3)_{0.06}$ target. The nickel and silver layers are 0.5 µm thick, while the YSZ is 2 µm thick.

The devices were measured using dilute hydrogen as a fuel (humidified Ar–4% H_2) and air as the oxidant. The maximum power outputs reported by the group were 6 mW cm^{-2} at 390 °C, the lowest temperature reported, and 145 mW cm^{-2} at 600 °C, the highest temperature. The stability of the power output and the device itself are not known, since the device was measured once, dwelling only 10 min at each temperature.

The electrodes used in the LLNL μSOFC are both single-phase, reducing the TPB length to just the perimeter of the electrode particles at the electrolyte interface. From Figure 4.7c and d, the electrode particle sizes appear to be roughly 300 nm–1 μm. Reducing these particle dimensions, or creating a porous composite of the materials with the electrolyte, would likely decrease the activation polarization. Since this is expected to be the dominant loss mechanism, improved electrodes should lead to enhanced device performance. Replacing the YSZ electrolyte with a doped ceria may improve device performance, though probably because ceria is more catalytically active at the electrode interfaces and not due to decreased electrolytic resistance.

Chen *et al.* from the University of Houston reported a μSOFC built on a 6 μm thick nickel foil substrate [36]. Pulsed laser deposition was used to deposit a YSZ film of about 2 μm thickness onto the foil. Photolithography was then used to define an etch pattern in the nickel foil, creating 70 μm diameter holes, separated by roughly 30 μm. To increase the TPB length, a 6 μm thick, porous nickel oxide–YSZ composite film was then deposited onto the etched nickel side of the structure, contacting the YSZ through the holes. Details regarding the deposition conditions or properties of this composite film were not provided. A 6 μm thick, porous $La_{0.5}Sr_{0.5}CoO_3$ (LSC) film was deposited by pulsed laser deposition onto the other side of the YSZ electrolyte. The maximum power output was 30 mW cm^{-2} at 480 °C and 110 mW cm^{-2} at 570 °C, similar to the LLNL device, with stable power output reported for 6 h.

There has been some research that suggests high electrochemical functionality can be derived from SOFC materials with high purity surfaces and nanometric grain sizes, as can be achieved with thin film processing routes [37]. In this work, yttria-stabilized zirconia and platinum thin films were deposited by sputtering and characterized in various configurations by impedance spectroscopy. The first configuration used dense platinum films micromachined into interdigitated electrodes on the surfaces of zirconia films as well as single crystal and polycrystalline bulk samples. A charge transfer resistance similar to what has been reported in the literature was measured for all of the bulk samples, and the films deposited at 600 °C. However, upon reducing the processing temperature below 600 °C, which can be done with vapor deposition but not in general with other methods of forming ceramic materials, a systematic decrease in the electrode resistance by a factor of nearly 1000 was discovered [38]. Using X-ray photoelectron spectroscopy, the mechanism of this enhancement was deduced to be the low-temperature films' increased surface purities. Such high surface purity is only possible when ceramics are produced at low temperatures, as can be done in vacuum environments.

The other configuration studied was a co-sputtering method to produce nanoscale composite platinum–YSZ films [39]. The stability of platinum and YSZ in both oxidizing and reducing environments allows this composite to be used as both anode and cathode. Using the same type of film for both electrodes reduces processing complexity and also creates a symmetric thermal expansion coefficient across the anode/electrolyte/cathode membrane. This latter benefit helps improve the membrane's thermomechanical stability by reducing bending stresses. Though further optimization of the composite microstructure is likely possible, the films exhibited high electrochemical functionality with expected device output of about $0.25\,W\,cm^{-2}$ at $600\,°C$.

Demonstrating the promise of this approach, a μSOFC device with high output power was reported by Huang et al., from Stanford University [40]. Structurally, the device is similar to the LLNL μSOFC, but uses porous, 80 nm thick, sputtered platinum films for both the anode and cathode. The electrolyte was a dense, sputtered YSZ film with thickness between 50 and 150 nm. A cross-sectional scanning electron micrograph of the device is given in Figure 4.8. The group reported significant power output at only $350\,°C$. Using hydrogen fuel, the peak power densities at this temperature were 60, 85 and $130\,mW\,cm^{-2}$ using electrolyte film thicknesses of 50, 100 and 150 nm, respectively. Even more impressive was the reported power output when the electrolyte was a bilayer with a 50 nm thick YSZ film on the anode side and a 50 nm thick gadolinium doped cerium oxide (GDC) film on the cathode side. Maximum power outputs of $200\,mW\,cm^{-2}$ at $350\,°C$ and $400\,mW\,cm^{-2}$ at $400\,°C$ were reported. These appear to be the highest ever reported for a SOFC operating below $400\,°C$.

Researchers at ETH have recently demonstrated a platform that takes advantage of the free-standing anode/electrolyte/cathode membrane to provide thermal insulation and thus reduce the power required to bring the device to operating temperature [41]. A diagram of the device is given in Figure 4.9. The membrane is made initially of silicon nitride, within which is buried a meandering platinum heater line. For this study, $La_{0.6}Sr_{0.4}Co_{0.2}Fe_{0.8}O_3$, a typical SOFC cathode material, was then deposited onto the membrane. The heater was then used to heat the membrane up to $800\,°C$ at rates of several $100\,°C\,min^{-1}$. To heat the device to

Figure 4.8 A scanning electron micrograph of a cross-section of the Stanford μSOFC. A porous Pt electrode is shown in the upper portion of the image. Reprinted from Ref. [40].

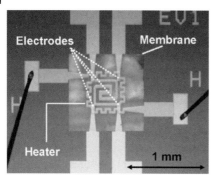

Figure 4.9 A scanning electron micrograph of a micro-hotplate fuel cell platform. Reprinted from Ref. [41].

600 °C, an expected operating temperature, required only 120 mW of input power. The functionality of the LSCF film on the membrane was demonstrated by the conductivity matching that of a film placed on a bulk substrate. Deposition of an electrolyte and an anode, as well as a means to etch holes in the supporting silicon nitride membrane, was not reported.

4.4
Polymer Electrolyte Membrane Fuel Cells

4.4.1
Basic Principles

Polymer fuel cells were invented in 1959 at General Electric [42]. As the name implies, this class of fuel cells uses one of a class of ion-conducting polymers for the electrolyte. These electrolytes conduct protons – that is hydrogen ions – from the anode to the cathode, and thus pure hydrogen is the fuel typically used. The conduction of protons leads to the alternate name for this type of fuel cell, the proton exchange membrane (both names yielding the acronym PEM). Polymeric electrolytes are sufficiently conductive to allow operation near room temperature. This is clearly an advantage for portable operation, since the issues of sealing, start-up and thermal containment are abated. On the other hand, such low temperatures inhibit the electrochemical reaction such that expensive platinum catalysts are required at the electrodes.

Typically, the electrolyte used is a perflourosulfonic acid membrane. Often, it is Nafion, a tetrafluoroethylene backbone copolymer with sulfonate-terminated side chains, manufactured by DuPont. Nafion and similar polymers are characterized by high chemical and thermal stability but also typically high cost. Conduction of ions occurs through, essentially, motion of hydrated protons along and between polymer chains. The stability and conductivity of polymer electrolytes is typically dependent upon hydration, such that the water management is a concern and, specifically, temperatures must be kept to less than 100 °C. Electrolyte hydration is also a concern because the membranes can change volume by the uptake and release of water. This problem is more severe when the membrane is thinner, but efforts have been made to minimize the issue through the use of composites [43].

The anode and cathode in a PEM are exposed to fuel and air, respectively, at approximately 90 °C, and so must be stable under these conditions. Traditionally, graphite is used as the backing layer of both the anode and cathode and platinum catalysts are dispersed across their surface to enhance the catalytic activities. The purpose of the backing layer is for mechanical support, to provide electronic conduction to the electrolyte/electrode interface, and to diffuse the reactant and waste streams to and from the electrolyte. Platinum loadings of $1\,mg\,cm^{-2}$ are regularly used; however, to reduce costs to levels competitive with other technologies, the Department of Energy has set a goal of maintaining similar levels of performance with platinum loadings of $0.4\,mg\,cm^{-2}$ [8]. Water management is an issue for the cathode, since the sub-100 °C operating temperature means water produced by the oxidation reaction at the cathode may be in liquid form. The cathode may become flooded by the waste water if it is not effectively removed.

Electrolyte and electrode materials that can perform at higher temperatures and with more complex fuels without degradation are of high interest. Increased operational temperature, if it could be handled by the electrolyte, would increase CO tolerance. Service lifetimes of 40 000 h are expected for stationary PEM fuel cell applications, though portable applications may require only 1000–5000 h [43]. Regardless, a very low degradation rate of the polymer electrolyte is required. While the platinum in the electrode can handle higher temperatures without failure, it is prone to poisoning from impurities in the fuel, particularly carbon monoxide. These impurities, even at part per million levels, can be irreversibly adsorbed on the platinum surface and block catalytic activity [8]. This places a heavy burden on the fuel reformer if a fuel other than pure H_2 is to be used.

4.4.2
Direct Methanol Fuel Cells

Depending on how it is stored, the volumetric energy density of hydrogen fuel can be very low and thus liquid fuels can be advantageous in many situations. Any hydrogen bearing fuel can, theoretically, be used with a PEM fuel cell if an external reformer first cracks off the hydrogen. However, for efficiency, it would be better to use the fuel directly without external reforming. Methanol is a promising candidate as a liquid fuel to use in fuel cells, and much research has been done to utilize methanol directly in polymer fuel cells. The chief concerns are poisoning of the anode catalyst and leakage of unreacted methanol through the electrolyte (i.e. crossover).

The electrolyte in a direct methanol fuel cell (DMFC) is typically the same proton conducting Nafion as used with hydrogen-fueled PEM cells. In presenting methanol in place of hydrogen gas to the anode, the electrode half-reaction must split and ionize the methanol so that protons are available to conduct through the electrolyte. Typically this requires that methanol reacts with water at the anode to oxidize the carbon from the methanol and liberate the hydrogen from the water and methanol. Advanced catalysts are required to cause this reaction to proceed and avoid carbon monoxide poisoning [44].

A crucial issue for DMFCs is that Nafion is highly permeable to methanol, which allows fuel to reach the cathode without first reacting electrochemically. Methanol is usually diluted with water to ameliorate these concerns; however, this reduces the overall energy density of the fuel. Water can be recycled from the cathode, though this adds to the complexity and likely reduces the efficiency of operation. Despite the increased complications of water management, the micro-DMFC is the most researched and likely closest to commercialization of the portable fuel cell technologies [45]. This probably is due to the reduced complexity compared to SOFCs given near-room temperature operation and the increased volumetric energy density compared to hydrogen-fueled PEM cells. Indeed, advanced portable DMFC power systems have been demonstrated [46–49]. Companies including Motorola, Samsung, Toshiba and others are believed to be developing this technology for portable electronics [50]. Research and some of the successful demonstrations will be discussed in the following section.

4.4.3
Micro Polymer Electrolyte Membrane Fuel Cells

Micro PEM fuel cells do not suffer from the high operation temperatures that go along with μSOFCs. This greatly reduces many of the processing and manufacturing difficulties in producing these fuel cells at the microscale (though authors have noted that even μPEM fuel cells will have thermal dissipation issues that are more complicated than for batteries [51]). There are a great number of research efforts, many proprietary, for optimization and, indeed, commercialization of μPEM fuel cells, including both hydrogen- and methanol-fueled devices. A number of good review articles are available [52–54].

The vast majority of μPEMs are produced by micromachining flow channels and current collectors on silicon wafers and then sandwiching these around conventional Nafion membranes loaded with carbon and platinum catalyst layers. This manufacturing route is possible because of the mechanical robustness of thin, polymeric Nafion membranes. Typically, the membrane and flow channels are held together by mechanical force with glue or gaskets used for sealing. Alternatively, the membrane can be formed on to a micromachined substrate by casting or spinning.

The chief difference between a PEM fabricated by this method as opposed to conventional routes is simply that the silicon wafer serves as both current collector and flow channel. This is a relatively small change; nevertheless, Yu *et al.* found that a PEM with a micromachined flow channel (shown schematically in Figure 4.10) significantly increased the power output at high currents [55]. This effect was correlated to higher reactant diffusivity to the electrodes compared with a conven-

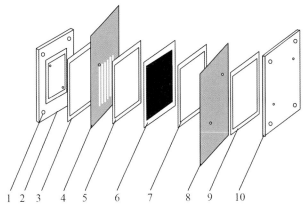

Figure 4.10 A diagram of the structure of a typical micro PEM fuel cell. Labeled parts represent: (1) assembly holes, (2, 10) end plates, (3, 5, 7, 9) gaskets, (4, 8) micromachined silicon wafers and (6) Nafion membrane with carbon/platinum catalyst. Reprinted from Ref. [55].

tional flow channel design. Power outputs from the µPEM, when operated on H_2 fuel, approached $200\,mW\,cm^{-2}$ at $25\,°C$ and were stable for over $300\,h$.

The fragility of silicon wafers has led some researchers to use other micromachined substrates for the current collector and flow plates. Lu *et al.* reported on the successful use of photochemically etched, thin stainless steel plates, with gold surface films to protect against corrosion [56]. The system was operated using methanol at dilutions of 1–4 M, with power outputs of $100\,mW\,cm^{-2}$ at $60\,°C$. However, methanol crossover at the gasket seals was noted. Substrates reported in other studies include glass [57], polymethyl methacrylate (PMMA) [58, 59] and polydimethylsiloxane (PDMS) [60].

While the thermal management issues of µSOFCs are largely avoided with polymer fuel cells, there are issues of water management to ensure stability and conductivity of the electrolyte as well as prevention of electrode flooding. Since energy density is likely to be a significant concern for micro fuel cell systems, fuel dilution is to be avoided as much as possible. Conventional water recycling methods such as compressors would add significantly to the size and weight of a micro fuel cell system, and would be difficult to manufacture. Thus, water management remains an area of high concern for commercialization of this technology.

Blum *et al.* reported on a direct methanol fuel cell that passively recycled water from the cathode back to the fuel [61]. The method used a $10–20\,µm$ thick layer of hydrophobic paste applied to the cathode current collector. This partially blocked liquid water from leaking out of the device and forced a portion to return to the anode due to hydraulic pressure. Water neutral operation (i.e. refueling with undiluted methanol) was found possible. The effects of the hydrophobic coating on power output were not fully explored.

4.5
New Concepts for Micro Fuel Cells

4.5.1
Single Chamber Fuel Cells

A number of researchers have reported power generation from fuel cells operating in the so-called single chamber or mixed reactant mode [62–68]. In this mode, the anode and cathode are placed in the same atmosphere, a mixture of the fuel and oxidant gases [62]. The chemical potential gradient that drives the fuel cell is created by the anode material being selectively catalytically active to the oxidation reaction and the cathode to the reduction reaction. This mode has the advantage of not requiring a seal between two gas environments. A single chamber fuel cell can thus be made planar, greatly easing the ability to miniaturize/microfabricate. However, it has a number of disadvantages, including: reduced V_{OC}, since the catalytic selectivities of the different electrodes are not perfect; reduced electrochemical oxidation of fuel, since direct chemical reaction in the gas phase may occur, and also very high flow rates are found necessary for reasonable power output [66]; and increased safety concerns, since fuel and oxidant are mixed (though typically in ratios outside the explosive range).

A possible issue with microfabricated single chamber fuel cells may arise when the anode and cathode are placed in close proximity. While such a configuration may be advantageous in reducing the ohmic polarization [13], it is not known to what degree this may affect the catalytic activity differential between the two electrodes required for single chamber operation. Whether by surface diffusion or gas phase transport, it is possible that semi-reacted species from either electrode may contact the opposite electrode and undergo direct chemical, as opposed to the desired electrochemical, reaction. Recent evidence suggests that this is indeed occurring [69]. The solution to this issue will require a cell geometry that reduces the transport of species between the electrodes except by way of ionic transport through the electrolyte. A number of possible solutions are detailed in Ref. [70].

While the single chamber fuel cell has a number of benefits for portable use, few efforts have been directed towards microfabricated single chamber fuel cells. An exception is the work of Shao et al., who reported a μSOFC operating in the single chamber mode [67]. This device was fabricated largely through powder processing methods and included a 0.7 mm thick, porous composite of nickel and Sm-doped ceria (SDC) as anode and substrate, a 20 μm thick SDC electrolyte and a 5–10 μm thick sprayed powder suspension of $Ba_{0.5}Sr_{0.5}Co_{0.8}Fe_{0.2}O_3$ (BSCF) as cathode [68]. A 10 μm thick Ru and pure ceria layer served as an anode catalyst.

The power output of the fuel cell was measured in a mixed propane/oxygen/He atmosphere. The most compelling feature of the fuel cell was that, after initiating cell operation in a furnace, the device could be removed from the external heat source and placed in thermal insulation. The device would then continue to operate as long as gas flow was maintained. In this thermally self-sustained mode, power outputs reached roughly $250\,mW\,cm^{-2}$. Electrolyte resistance was found to be responsible for more than half of the total polarization, so the power output may be improved significantly using a thinner electrolyte.

Whether the device could be portable is debatable. The processing methods are not typically considered microfabrication. Also, since the thick substrate served as the anode, there is a large thermal mass that must be heated during initialization and then maintained at operating temperature. A supplementary battery to provide the start-up energy and the insulation used to contain the heat would add significantly to the size and weight of a packaged device. Finally, fuel utilization was estimated to be only 1%. This must be improved significantly before the device is competitive with the energy densities of existing battery solutions.

4.5.2
Membraneless (Liquid Laminar Flow) Micro Fuel Cell

A new fuel cell design uniquely suited for operation in a microfluidic device is the membraneless fuel cell. This design uses the laminar flow regime possible in a microfluidic cell to operate without a traditional "membrane" electrolyte. One entrant liquid stream contains the fuel while the other contains an oxidant. The two streams flow in parallel down a channel and the only "membrane" is the liquid–liquid interface. The fuel, often formic acid, and the oxidant, often oxygen, are dissolved in water. The electrodes are placed on opposite sides of the flow channel, and the ionic transport occurs by diffusion of protons through the aqueous reactant streams. To improve the conductivity of this "electrolyte," acids can be added to the fuel and oxidant solutions, thus increasing the proton concentration. A diagram of such a device is given in Figure 4.11.

The benefit of this arrangement is removal of all of the issues associated with membrane stability. This new type of fuel cell has been extensively researched by

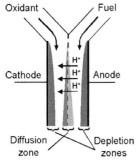

Figure 4.11 Schematic of the membraneless liquid laminar flow fuel cell (not drawn to scale). The dashed line represents the flow boundary between the fuel and oxidant streams; no physical membrane separates the two. Reprinted from Ref. [71].

a group at the University of Illinois [71–74], in addition to others [75, 76]. Low solubility and low diffusivity of the reactants within the reactant streams can limit the power outputs of the cells. To overcome this, one or both of the electrodes have been converted to gas breathing, while still maintaining a liquid-phase electrolyte. Power outputs approaching $200\,mW\,cm^{-2}$ at room temperature have been reported in this configuration [74], with open circuit voltages equal to that reported by traditional PEM fuel cells.

4.6
Conclusions

The high conversion efficiencies, high energy densities and relatively simple operation of fuel cells is a strong motivation for their use as micro chemical power generators. The use of microfabrication techniques is promising for device miniaturization and improvement of power output density. The improvements may include decreased concentration polarization from microfabricated reactant flow channels and decreased ohmic polarization from thin electrolytes. On the other hand, activation polarization remains a difficult problem for thin electrodes. Improvements may be found in this regard, however, with nanometric structuring of the electrodes, non-planar substrate geometries or employment of multilayer manufacturing techniques.

Significant engineering hurdles remain for these devices, specifically in the integration and microfabrication of materials not commonly used in microsystems, and in areas of thermal and chemical management. The number of research efforts directed towards micro fuel cells, including those in industry, suggests a general optimism that solutions to these obstacles will be found.

Acknowledgements

The support of the DoD Multidisciplinary University Research Initiative (MURI) program administered by the Army Research Office under Grant No. DAAD19-01-1-0566 and the National Science Foundation under Grant No. DMR-0243993 are appreciated.

References

1 Grove, W.R. (1839) On voltaic series and the combination of gases by platinum. *Philosophical Magazine*, **14**, 127–30.

2 Cowey, K., Green, K.J., Mepsted, G.O. and Reeve, R. (2004) Portable and military fuel cells. *Current Opinion Solid State and Materials Science*, **8**, 367–71.

3 Mason, J. (2004) Micro fuel cells headed to market, and a showdown. *Small Times*, 27 May 2004.

4 Srinivasan, S., Mosdale, R., Stevens, P. and Yang, C. (1999) Fuel cells: reaching the era of clean and efficient power generation in the twenty-first century.

Annual Review of Energy and Environment, **24**, 281–328.

5 Nowotny, J., Bak, T., Nowotny, M.K. and Sorrell, C.C. (2005) Charge transfer at oxygen/zirconia interface at elevated temperatures, part 2: oxidation of zirconia. *Advances in Applied Ceramics,* **104**, 154–64.

6 Bard, A.J., Faulkner, L.R. and Methods, E. (2004) *Fundamentals and Applications,* 2nd edn, John Wiley & Sons, Singapore.

7 Adler, S.B. (2004) Factors governing oxygen reduction in solid oxide fuel cell cathodes. *Chemical Reviews,* **104**, 4791–843.

8 United States Department of Energy (2002) *Fuel Cell Handbook,* 6th edn, United States Department of Energy, Morgantown, WV.

9 Tuller, H.L. (1981) Mixed conduction in nonstoichiometric oxides, in *Nonstoichiometric Oxides* (ed. O.T. Sorensen), Academic Press, New York, pp. 271–335.

10 Steele, B.C.H. and Heinzel, A. (2001) Materials for fuel-cell technologies. *Nature,* **414**, 345–52.

11 Charpentier, P., Fragnaud, P., Schleich, D.M. and Gehain, E. (2000) Preparation of thin film SOFCs working at reduced temperature. *Solid State Ionics,* **135**, 373–80.

12 de Souza, S., Visco, S.J. and De Jonghe, L.C. (1997) Reduced-temperature solid oxide fuel cell based on YSZ thin-film electrolyte. *Journal of the Electrochemical Society,* **144**, L35–7.

13 Fleig, J., Tuller, H.L. and Maier, J. (2004) Electrodes and electrolytes in micro-SOFCs: a discussion of geometrical constraints. *Solid State Ionics,* **174**, 261–70.

14 Chachuat, B., Mitsos, A. and Barton, P.I. (2005) Optimal design and steady-state operation of micro power generation employing fuel cells. *Chemical Engineering Science,* **60**, 4535–56.

15 Xu, X., Xia, C., Huang, S. and Peng, D. (2005) YSZ thin films deposited by spin-coating for IT-SOFCs. *Ceramics International,* **31**, 1061–4.

16 Mogensen, M., Primdahl, S., Jørgensen, M.J. and Bagger, C. (2000) Composite electrodes in solid oxide fuel cells and

similar solid state devices. *Journal of Electroceramics,* **5**, 141–52.

17 Srikar, V.T., Turner, K.T., Ie, T.Y.A. and Spearing, S.M. (2004) Structural design considerations for micromachined solid-oxide fuel cells. *Journal of Power Sources,* **125**, 62–9.

18 Solasi, R., Zou, Y., Huang, X., Reifsnider, K. and Condit, D. (2007) On mechanical behavior and in-plane modeling of constrained PEM fuel cell membranes subjected to hydration and temperature cycles. *Journal of Power Sources,* **167**, 366–77.

19 Tang, Y., Stanley, K., Wu, J., Ghosh, D. and Zhang, J. (2005) Design considerations of micro thin film solid-oxide fuel cells. *Journal of Micromechanics and Microengineering,* **15**, S185–192.

20 Baur, E. and Preis, H. (1937) Über Brennstoff-Ketten mit Festleitern. *Zeitschrift fur Elektrochemie,* **43**, 727–32.

21 Ralph, J.M., Schoeler, A.C. and Krumpelt, M. (2001) Materials for lower temperature solid oxide fuel cells. *Journal of Materials Science,* **36**, 1161–72.

22 Minh, N.Q. (1993) Ceramic fuel cells. *Journal of the American Ceramic Society,* **76**, 563–88.

23 Mogensen, M. and Kammer, K. (2003) Conversion of hydrocarbons in solid oxide fuel cells. *Annual Review of Materials Research,* **33**, 321–31.

24 Goodenough, J.B. (2003) Oxide-ion electrolytes. *Annual Review of Materials Research,* **33**, 91–128.

25 Larminie, J. and Dicks, A. (2003) *Fuel Cell Systems Explained,* 2nd edn, John Wiley & Sons, Ltd, Chichester.

26 Arana, L.R., Schaevitz, S.B., Franz, A.J., Schmidt, M.A. and Jensen, K.F. (2003) A microfabricated suspended-tube chemical reactor for thermally efficient fuel processing. *Journal of Microelectromechanical Systems,* **12**, 600–12.

27 Baertsch, C.D., Jensen, K.F., Hertz, J.L., Tuller, H.L., Vengallatore, S.T., Spearing, S.M. and Schmidt, M.A. (2004) Fabrication and structural characterization of self-supporting electrolyte membranes for a micro-solid oxide fuel cell. *Journal of Materials Research,* **19**, 2604.

28 Bruschi, P., Diligenti, A., Nannini, A. and Piotto, M. (1999) Technology of integrable

free-standing yttria-stabilized zirconia membranes. *Thin Solid Films*, **346**, 251–4.

29 Nair, J.P., Wachtel, E., Lubomirsky, I., Fleig, J. and Maier, J. (2003) Anomalous expansion of CeO2 nanocrystalline membranes. *Advanced Materials*, **15**, 2077–81.

30 Jankowski, A.F., Hayes, J.P., Graff, R.T. and Morse, J.D. (1999) Testing of solid oxide fuel cells for micro to macro power generation, in *Solid Oxide Fuel Cells VI* (eds M. Dokiya and S. Singhal), Electrochemical Society Proceedings, Pennington, PA, pp. 932–7.

31 Jankowski, A.F., Hayes, J.P., Graff, R.T. and Morse, J.D. (2002) Micro-fabricated thin-film fuel cells for portable power requirements, in *Materials for Energy Storage, Generation and Transport* (eds R.B. Schwarz, G. Ceder and S.A. Ringel), Materials Research Society Proceedings, Warrendale, PA, pp. V4.2.1–6.

32 Morse, J. and Jankowski, A. (2002) A MEMS-based fuel cell for microscale energy conversion. (Lawrence Livermore National Laboratory Technical Reports, Livermore, CA). http://www.llnl.gov/tid/lof/documents/pdf/243986.pdf (accessed 24 September 2007).

33 Jankowski, A.F. and Morse, J.D. MEMS-based thin-film fuel cells. United States Patent 6,638,654, 1 February 1999.

34 Company and product information found at http://www.ultracellpower.com (accessed 24 September 2007).

35 Jankowski, A.F. and Hayes, J.P. (2002) Sputter deposition of metallic sponges. *Journal of Vacuum Science and Technology A*, **21**, 422–5.

36 Chen, X., Wu, N.J., Smith, L. and Ignatiev, A. (2004) Thin-film heterostructure solid oxide fuel cells. *Applied Physics Letters*, **84**, 2700–2.

37 Hertz, J.L. (2006) Microfabrication methods to improve the performance of the yttria-stabilized zirconia – platinum – oxygen electrode, PhD Thesis, Massachusetts Institute of Technology.

38 Hertz, J.L., Rothschild, A. and Tuller, H.L. (2002) Highly enhanced electrochemical performance of silicon-free platinum/yttria stabilized zirconia

interfaces. *Journal of Electroceramics*, DOI: 10.1007/S10832-008-9475-5 (15 April 2008).

39 Hertz, J.L. and Tuller, H.L. (2007) Nanocomposite platinum-yttria stabilized zirconia electrode and implications for micro solid oxide fuel cell operation. *Journal of the Electrochemical Society*, **154**, B413.

40 Huang, H., Nakamura, M., Su, P., Fasching, R., Saito, Y. and Prinz, F.B. (2007) High-performance ultrathin solid oxide fuel cells for low-temperature operation. *Journal of the Electrochemical Society*, **154**, B20–24.

41 Beckel, D., Briand, D., Bieberle-Hütter, A., Courbat, J., de Rooij, N.F. and Gauckler, L.J. (2007) Micro-hotplates – a platform for micro-solid oxide fuel cells. *Journal of Power Sources*, **166**, 143–8.

42 Grubb, W.T. Fuel Cell. US Patent 2,913,511, 17 Nov. 1959.

43 Rajendran, R.G. (2005) Polymer electrolyte membrane technology for fuel cells. *MRS Bulletin/Materials Research Society*, **30**, 587–90.

44 Wasmus, S. and Kuver, A. (1999) Methanol oxidation and direct methanol fuel cells: a selective review. *Journal of Electroanalytical Chemistry*, **461**, 14–31.

45 Morse, J.D. (2007) Micro-fuel cell power sources. *International Journal of Energy Research*, **31**, 576–602.

46 Dillon, R., Srinivasan, S., Arico, A.S. and Anotonucci, V. (2004) International activities in DMFC R&D: status of technologies and potential applications. *Journal of Power Sources*, **127**, 112–26.

47 Ren, X.M., Zelenay, P., Thomas, S., Davey, J. and Gottesfeld, S. (2000) Recent advances in direct methanol fuel cells at Los Alamos National Laboratory. *Journal of Power Sources*, **86**, 111–16.

48 Kim, D.J., Cho, E.A., Hong, S.A., Oh, I.H. and Ha, H.Y. (2004) Recent progress in passive direct methanol fuel cells at KIST. *Journal of Power Sources*, **130**, 172–7.

49 Yoshitake, T., Kimura, H., Kuroshima, S., Watanabe, S., Shimakawa, Y., Manako, T., Nakamura, S. and Kubo, Y. (2002) Small direct methanol fuel cell pack for portable applications. *Electrochemistry*, **70**, 966–8.

50 Oedegaard, A. and Hentschel, C. (2006) Characterisation of a portable DMFC stack

and a methanol-feeding concept. *Journal of Power Sources*, **158**, 177–87.

51 Meyers, J.P. and Maynard, H.L. (2002) Design considerations for miniaturized PEM fuel cells. *Journal of Power Sources*, **109**, 76–88.

52 Nguyen, N.T. and Chan, S.H. (2006) Micromachined polymer electrolyte membrane and direct methanol fuel cells – a review. *Journal of Micromechanics and Microengineering*, **16**, R1–12.

53 Morse, J.D. (2007) Micro-fuel cell power sources. *International Journal of Energy Research*, **31**, 576–602.

54 Kundu, A., Jang, J.H., Gil, J.H., Jung, C.R., Lee, H.R., Kim, S.-H., Ku, B. and Oh, Y.S. (2007) Micro-fuel cells – current development and applications. *Journal of Power Sources*, **170**, 67–78.

55 Yu, J., Cheng, P., Ma, Z. and Yi, B. (2003) Fabrication of miniature silicon wafer fuel cells with improved performance. *Journal of Power Sources*, **124**, 40–6.

56 Lu, G.Q. and Wang, C.Y. (2005) Development of micro direct methanol fuel cells for high power applications. *Journal of Power Sources*, **144**, 141–5.

57 Lee, S.J., Chang-Chien, A., Cha, S.W., O'Hayre, R., Park, Y.I., Saito, Y. and Prinz, F.B. (2002) Design and fabrication of a microfuel cell array with "flip-flop" interconnection. *Journal of Power Sources*, **112**, 410–18.

58 Hsieh, S.S., Kuo, J.K., Hwang, C.F. and Tsai, H.H. (2004) A novel design and microfabrication for a micro PEMFC. *Microsystems Technologies*, **10**, 121–6.

59 Chan, S.H., Nguyen, N.-T., Xia, Z. and Wu, Z. (2005) Development of a polymeric micro fuel cell containing laser-micromachined flow channels. *Journal of Micromechanics and Microengineering*, **15**, 231–6.

60 Shah, K., Shin, W.C. and Besser, R.S. (2003) A PDMS micro proton exchange membrane fuel cell by conventional and non-conventional microfabrication techniques. *Sensors and Actuators B*, **97**, 157–67.

61 Blum, A., Duvdevani, T., Philosoph, M., Rudoy, N. and Peled, E. (2003) Water-neutral micro direct-methanol fuel cell (DMFC) for portable applications. *Journal of Power Sources*, **117**, 22–5.

62 Dyer, C.K. (1990) A novel thin-film electrochemical device for energy conversion. *Nature*, **343**, 547–8.

63 Hibino, T., Hashimoto, A., Inoue, T., Tokuno, J., Yoshida, S. and Sano, M. (2000) A low-operating temperature solid oxide fuel cell in hydrocarbon-air mixtures. *Science*, **288**, 2031–3.

64 Napporn, T.W., Morin, F. and Meunier, M. (2004) Evaluation of the actual working temperature of a single-chamber SOFC. *Electrochemical and Solid-State Letters*, **7**, A60–2.

65 Buergler, B.E., Siegrist, M.E. and Gauckler, L.J. (2005) Single chamber solid oxide fuel cells with integrated current-collectors. *Solid State Ionics*, **176**, 1717–22.

66 Stefan, I.C., Jacobson, C.P., Visco, S.J. and De Jonghe, L.C. (2004) Single chamber fuel cells: flow geometry, rate, and composition considerations. *Electrochemical and Solid-State Letters*, **7**, A198–200.

67 Shao, Z., Haile, S.M., Ahn, J., Ronney, P.D., Zhan, Z. and Barnett, S.A. (2005) A thermally self-sustained micro solid-oxide fuel-cell stack with high power density. *Nature*, **435**, 795–8.

68 Shao, Z. and Haile, S.M. (2004) A high-performance cathode for the next generation of solid-oxide fuel cells. *Nature*, **431**, 170–3.

69 Ahn, S.J., Kim, Y.B., Moon, J., Lee, J.H. and Kim, J. (2007) Influence of patterned electrode geometry on performance of co-planar, single-chamber, solid oxide fuel cell. *Journal of Power Sources*, **171**, 511–16.

70 Hertz, J.L. and Tuller, H.L. Micro fuel cell. U.S. Patent Application 20070141445, 21 June 2007.

71 Choban, E.R., Markoski, L.J., Wieckowski, A. and Kenis, P.J.A. (2004) Microfluidic fuel cell based on laminar flow. *Journal of Power Sources*, **128**, 54–60.

72 Choban, E.R., Waszczuk, P. and Kenis, P.J.A. (2005) Characterization of limiting factors in laminar flow-based membraneless microfuel cells. *Electrochemical and Solid-State Letters*, **8**, A348–52.

73 Jayashree, R.S., Gancs, L., Choban, E.R., Primak, A., Natarajan, D., Markoski, L.J. and Kenis, P.J.A. (2005) Air-breathing laminar flow based microfluidic fuel cell. *Journal of the American Chemical Society*, **127**, 16758–9.

74 Jayashree, R.S., Mitchell, M., Natarajan, D., Markoski, L.J. and Kenis, P.J.A. (2007) Microfluidic hydrogen fuel cell with a liquid electrolyte. *Langmuir*, **23**, 6871–4.

75 Cohen, J.L., Westly, D.A., Pechenik, A. and Abruna, H.D. (2005) Fabrication and preliminary testing of a planar membraneless microchannel fuel cell. *Journal of Power Sources*, **139**, 96–105.

76 Bazylak, A., Sinton, D. and Djilali, N. (2005) Improved fuel utilization in microfluidic fuel cells: a computational study. *Journal of Power Sources*, **143**, 57–66.

5
Microscale Heat Engines

Stuart Jacobson, Hanqing Li and Alan Epstein

5.1
Introduction

Heat engines are the power conversion choice for most large-scale power plants. At the megawatt level, coal and nuclear power boil pressurized water to propel steam turbine-driven electric generators. From the megawatt to kilowatt level, natural gas heats pressurized air to propel gas turbine-driven electric generators. At the kilowatt level, gasoline-fueled (internal combustion) electric generators are available as electric back-up for home use. These heat engine examples all contain dynamic components moving at high speeds to obtain desirable levels of conversion efficiency. The vast majority of electric power used in the world today comes from heat engines of some sort. However, for portable applications, batteries have been the primary electric power source. With increasing need for improved portable power sources, could heat engines be an option? This section explores ongoing research and development efforts aimed at answering this question.

In comparison to a battery, the compelling reasons for a microscale heat engine are energy and power density. The significance of energy density has been covered earlier in this book. The microscale heat engines discussed in this section burn hydrocarbon fuels containing chemical energy density that is about two orders of magnitude larger than that available from lithium-ion rechargeable batteries. Heat engines are relatively flexible in terms of the fuels they can burn; a wide range of hydrocarbon fuels and hydrogen are acceptable. While the second law of thermodynamics limits the conversion efficiency of a heat engine, overall conversion efficiencies (chemical-to-electric) of 10–20% should be feasible, resulting in a fuel energy density that is about an order of magnitude larger than a lithium-ion battery.

A proper comparison between a battery and a fueled heat engine must include the mass of the engine and fuel tank along with the fuel when calculating energy density. Liquid hydrocarbons such as propane are the fuels of choice to keep fuel (mass) density high, minimizing fuel tank size and mass. Engine size and mass are driven by power density, the ratio of engine power to engine size or mass. Power through a heat engine scales with mass flow rate, which scales with flow

Microfabricated Power Generation Devices. Edited by Alexander Mitsos and Paul I. Barton
Copyright © 2009 WILEY-VCH Verlag GmbH & Co. KGaA, Weinheim
ISBN: 978-3-527-32081-3

passage area. Engine mass and size scales with volume. So power density will scale like the ratio of area to volume, a ratio that simplifies to the inverse of length scale. Thus, with all else unchanged, as a heat engine is reduced in size, power density will increase in direct proportion to the length scale decrease. This scaling suggests that small-scale engines will have even larger power density than conventional-scale heat engines, which already have high power density (thus their choice for airplanes and cars).

Of course, all else is not unchanged, and the performance of a microscale engine will generally not be as good as a conventional-scale engine due to both physical, fabrication and material constraints. However, analysis has shown that the power density of a microscale engine can be quite good. For the MIT microscale gas turbine engine discussed later in this section, models indicate that an engine the size of a large button (about $2\,cm^3$) together with some power electronics (also the size of a button) is sufficient to produce $10\,W$ electric. The mass of this engine, of order a few grams, is small enough to be negligible in comparison to the fuel (and fuel tank) mass for any application requiring more than a few minutes of operation. In general, engine mass will depend on the heat engine cycle and performance.

Several microscale heat engine efforts have blossomed over the last decade. The focus of this chapter, in keeping with this book, will be on engines fabricated using silicon-based microfabrication processes. Small-scale engines fabricated using conventional machining techniques are beyond the scope of this chapter, although brief mention will be made of a few such efforts.

5.2
Challenges

Conventional-scale engines can achieve overall chemical-to-electric efficiencies of the order of 25–60%, depending on cycle and design. Efficiency for microscale engines is currently limited by physical, fabrication and material constraints. Assuming current (2007) state-of-the-art microfabrication capabilities and materials, efficiency levels in the 10–20% range for microscale heat engines are a significant challenge. However, future technological innovations could alleviate the fabrication and material constraints, improving the prospects for higher efficiency. This section will discuss the challenges associated with microscale engines, with a bias towards their impact on a gas turbine implementation, the field in which the authors of this chapter work.

5.2.1
Physical Challenges

To obtain high power densities, microscale engines need to operate in velocity regimes comparable to conventional-scale engines. However, due to their small size, boundary layers in microscale engines will clearly be much thinner than in conventional-scale engines simply because there is not a lot of distance over which the boundary layers can grow. This phenomenon is described physically by the dimensionless parameter, the Reynolds number, which scales linearly with length scale, and will be relatively small for microscale engines, indicating laminar/transitional flow for the most part. A thin boundary layer implies relatively large viscous losses. A thin thermal boundary layer implies relatively large heat transfer. Both of these phenomena have adverse implications for the performance of microscale engines.

5.2.1.1 Boundary Layer Effects

Jacobson [1] computed the effects of low Reynolds number on turbomachinery flows, showing that as Reynolds number (length scale) decreases, pressure loss increases and efficiency decreases. However, the results also indicate that these effects are moderate down to a certain device size, beyond which they begin to increase substantially. For instance, scaling a centrifugal compressor down to 1 cm in diameter cost only a few points in efficiency; however, efficiency decreased rapidly as the size was further reduced. By 4 mm in diameter, the compressor efficiency was no longer acceptable for use in a gas turbine cycle. The turbine showed similar scaling effects when operated cold. However, for gas turbine engine operation, the flow through the turbine is hot, resulting in lower fluid mass density and thus smaller Reynolds number, meaning that a hot turbine behaves like a much smaller cold turbine. Computations indicated that the performance of an 8 mm diameter turbine is already close to the limits of acceptability for a gas turbine cycle, implying that microscale gas turbines much smaller than this size are unlikely to be viable. The loss discussed here is focused on turbomachinery, but the increase in viscous loss at small scale impacts all microscale heat engines to some degree.

While boundary layers will be thin, the relative size of the boundary layer with respect to flow passage size actually increases as the length scale decreases. In a laminar boundary layer with constant free stream velocity, the boundary layer thickness is proportional to the square root of the development length. Since the development length scales with length scale, the ratio of boundary layer thickness to passage size scales like the inverse square root of length scale; as the length scale decreases, this ratio increases. The ratio increase holds true when comparing microscale laminar boundary layers to larger scale turbulent boundary layers as well, although the scaling is somewhat weaker.

This scaling effect can be important in flow regions that are not fully developed, such as along turbomachinery blades. Boundary layers on blades cause flow blockage, decreasing mass flow for a given passage size. At the microscale, the relatively thicker boundary layers result in more flow blockage, decreasing the relative mass flow. Thus power will not scale directly with flow area, and the power density gain from scale reduction will not be quite as good as the scaling described earlier. In decelerating flows, such as compressors, boundary layers grow more quickly, reducing the degree of this scaling effect. In accelerating flows, such as turbines, this scaling effect becomes more pronounced.

5.2.1.2 Heat Transfer Effects

Thermal boundary layers are thinner on average in microscale heat engines, resulting in larger heat transfer per unit area. Since power scales approximately like area, heat loss to the walls relative to the combustion power generated in the device will also increase. So microscale engines will lose relatively more heat to their walls than conventional-scale engines.

Heat transferred into a wall can act differently at the microscale. The Biot number is a dimensionless parameter defined as the ratio of convective heat transfer into a wall to heat conduction through the wall. Biot number scaling can be quantified for constant free stream velocity flow along a constant temperature wall. Convective heat transfer per unit area into the wall scales with the inverse square root of the length scale. Conduction heat transfer per unit area through the wall scales with the inverse of the length scale. So Biot number scales with the square root of the length scale, decreasing as the length scale decreases. The implication is that, as length scale decreases, the rate at which heat spreads through a structure rises more quickly than the rate at which heat is convected into that structure from the surrounding fluid. Thus, temperature gradients within microscale structures can be considerably smaller than in conventional-scale structures, making microscale structures appear to be nearly isothermal.

Thus the relative heat loss increase to the walls discussed earlier has the potential to do double damage to microscale engine performance. In a direct manner, heat loss to the walls decreases the combustor efficiency, decreasing engine performance, but the heat loss to the walls has the potential to do far more damage to the engine cycle resulting from the Biot number scaling effect. In a microscale gas turbine engine, as the compressor wall temperature increases from heat loss into the combustor wall, more heat is transferred into the compressor flow, increasing the power required for the compression process and thus reducing the excess turbine power available to drive the electric generator. This secondary effect of combustor heat loss has the potential to decrease engine performance significantly.

The reduction in Biot number also has a positive effect. Highly stressed structural components in the hot part of a microscale heat engine, such as the turbine blades in a gas turbine, may not need to be actively cooled as they would be, at considerable cost, in a conventional-scale engine. At the microscale, the heat transferred into the highly stressed components is conducted away to the cooler sections of the engine, keeping the structure temperature down, reducing thermal stress concerns.

5.2.2
Fabrication and Material Challenges

5.2.2.1 Standard Silicon Microfabrication

So long as the engine scale is sufficiently large (diameter of the order of 1 cm), many of the performance limitations of microscale heat engines can be traced back to limitations associated with the microfabrication tool set. The state of the art in microfabrication capabilities and materials is quite limited as this field is just emerging from its infancy. Several of these limitations are discussed below as they pertain to microscale engines but it must be kept in mind that these limitations are not necessarily inherent to the microscale. With significant investment in microfabrication process research and tool development, it is conceivable that the challenges discussed in this section could be greatly diminished or even eliminated.

Micro-electro-mechanical systems (MEMS) fabrication currently has relatively low precision. Microfabrication allows one to build complex components at very small length scales, but this is a separate issue from precision. At the macro scale, conventional machining tools can routinely produce meter-scale components with tolerances of the order of $10\,\mu m$, resulting in about five orders of magnitude difference between the smallest and the largest scales. With microfabrication, the largest components are centimeter-scale, whereas etch tolerance, when etching to depths of tens of μm, and mask and wafer alignment tolerances are all at best of the order of $1–5\,\mu m$, resulting in about four orders of magnitude difference between the smallest and largest scales.

This reduction in precision impacts one's ability to design components that need to fit together precisely or need to move relative to each other with small gaps between them, for example seals and bearings. For seals, as gap size increases, there is more leakage as a percentage of engine mass flow rate, resulting in a decrease in engine performance.

Standard photolithography is primarily a two-dimensional process. A two-dimensional pattern is transferred into photoresist, which is used to mask off parts of the wafer during a bulk etch. The resulting silicon structures are then simply extensions of the initial two-dimensional mask design. The ability to vary etch depth is very limited when etching to depths of tens or hundreds of μm.

These limitations in the types of three-dimensional structures that can be fabricated severely impact engine performance. For instance, in a typical centrifugal compressor design, blade height decreases as radius increases to compensate for the increase in circumferential distance, allowing one to control flow area change as a function of radius. However, standard microfabricated centrifugal compressor blades have constant blade height, so flow area cannot be controlled by blade height manipulation as the radius increases, making diffusion control a challenge, resulting in blades that are highly prone to separation. The resulting increase in adverse pressure gradient experienced as flow passes through such a microscale compressor results in thicker boundary layers, leading to increased blockage, reduced mass flow, and thus reduced power density. The Ghodssi group at the University of Maryland [2, 3] are developing gray scale lithography techniques that will provide a significant increase in one's ability to vary etch depth, thus helping to alleviate this issue.

Standard MEMS microfabrication has a limited material set. Currently, silicon is the only bulk material available that can be precision etched to depths of hundreds of μm. Single crystal silicon wafers are relatively inexpensive, although the solar cell industry has recently demanded a substantial fraction of the low thickness variation wafers, driving up prices. Single crystal silicon is an excellent mechanical material, with yield strength measured in excess of several GPa [4] and a density one third of that of the nickel superalloys used in large gas turbines. While silicon is a brittle material at room temperature, it becomes elastic at the elevated temperatures typical of engines. However, silicon has two main challenges for use in engines of this size. First, silicon has a very high thermal conductivity, higher than steel and just a bit lower than copper. This high thermal conductivity contributes to the very low Biot number discussed above, which makes it difficult to thermally isolate sections of an engine. Second, silicon begins to exhibit creep behavior under moderate stress levels at temperatures above 900 K [5], similar to nickel superalloys. Creep is a significant life issue for highly stressed components in hot engine sections such as turbine blades and disks.

Notwithstanding the constraints already discussed, there are significant advantages to microfabrication when designing very small components. Microfabrication allows one to work at length scales that are not achievable with standard tools. In addition, when designing patterns into masks, complexity in the mask plane comes for free. There is essentially no cost difference between simple and complex pattern designs on a mask and there is no real cost difference in etching those patterns into a wafer. So the designer has considerable leeway in the patterns that can be designed into the mask. The key constraint, as discussed above, comes from out-of-plane complexity.

One of the major benefits of microfabrication is the ability to batch fabricate; one typically builds multiple components (dies) on each silicon wafer. Many devices are fabricated in parallel during each pass through a processing tool. For instance, Figure 5.1 shows a silicon wafer containing an array of compressors. Once a fabrication process is understood and proper controls are in place, the process can be repeated with nearly identical results. As proven by the integrated circuit industry, batch fabrication with robust processes can lead to low cost components, so long as a large number of devices are being fabricated. Extrapolating costs from integrated circuits to MEMS must be done with care as the processes and volumes are quite different. However, the success of the integrated circuit

Figure 5.1 An array of compressors etched into a silicon wafer, fabricated at MIT for a turbocharger device.

industry suggests that microfabricated heat engines could have a significant cost advantage over small-scale conventionally machined engines.

5.2.2.2 Small-Scale Conventional Machining

Small-scale variants of conventional machining tools, such as micro-milling machines are available for making centimeter-scale engines. Several research groups are pursuing this path, and their work will be briefly mentioned in the next section. This fabrication method has the great advantage of opening the space of materials that can be used within the engine, and allowing for components that can be truly three-dimensional. The difficulties in pursuing this path include realizing the fractional precision levels needed and controlling the cost of fabricating and assembling the engines.

The current state-of-the-art of microfabrication technology greatly limits design complexity, which limits engine performance. On the other hand, micro-milling engine components can lead to cost and assembly issues. It is too soon to say whether one path is preferable to the other and it may well be that a hybridized scheme that mixes both types of fabrication is the best route to pursue, at least in the near term.

5.3
Microscale Heat Engine Examples

In this section, several examples of silicon-based microscale heat engines will be described. Other small-scale non-silicon engines will be touched on for comparison but they will not be discussed in detail. The first major research program in this field was the microscale gas turbine engine project developed at MIT by Epstein and Senturia [6] starting in the mid-1990s. Other efforts followed shortly thereafter, developing internal combustion engines and steam engines. Richards and his group at Washington State University are pursuing a unique dynamic heat engine design utilizing two-phase flow density differences to periodically flex a piezoelectric membrane. This section will review several of these efforts, focusing on the MIT and Washington State research programs, as these are among the furthest advanced and illustrate the range of possibilities offered at this scale.

5.3.1
MIT MEMS Gas Turbine Generator

MIT is leading an effort to develop microfabricated gas turbine generators for portable power applications [7]. These devices would deliver 10–20 W of electric power from a 5 cm^3 engine with a mass of about 10 g (including power electronics, but excluding fuel and fuel tank). The initial goal of this program is to obtain engines capable of achieving 5% chemical-to-electric conversion efficiency. Later design improvements would allow an efficiency increase to at least 10%.

The MIT group developed several design guidelines for the fabrication of such a complex MEMS device. First, to maintain consistency with batch fabrication, all processes have minimal human intervention at the die scale. Second, most processes are performed at the individual wafer level, prior to bonding with other wafers. This guideline has driven up device yield, since an out-of-spec process performed on a bonded wafer stack can result in considerable loss of time and effort. Third, new devices evolve from previously demonstrated devices, utilizing proven robust process steps and geometrical configurations where possible. This guideline improves fabrication repeatability and helps isolate operational issues during testing.

One of the key challenges in developing a new MEMS device is the difficulty in making changes to the device once it is built. At the macroscale, one can generally disassemble and modify components of a device being fabricated. At the microscale, once an out-of-spec component is identified in a bonded stack, there is little one can do but start again. The guidelines discussed above increased yield and pushed the development effort along. However, at the same time, the geometric and programmatic challenges associated with microfabrication remain the largest barriers to device development, far more so than physical limitations at the microscale. As a result, the microscale gas turbine engine design is fundamentally a trade-off between fabrication complexity and engine functionality/performance.

The MIT program initially demonstrated critical technologies required for gas turbine operation, including bearings, combustion, turbomachinery and electric generation. Later work focused on integrating these technologies into an engine.

5.3.1.1 Bearings and Rotordynamics

Gas turbine engines require transonic rotational speeds to achieve high levels of turbomachinery performance. Considerable effort was devoted to developing gas bearing and rotordynamic models consistent with the microscale design space and fabrication constraints and several devices were developed to verify these models. One of the first devices built in the MIT program was a high-speed rotor demonstrator consisting of a bonded stack of five silicon wafers. The middle wafer contained a 4.2 mm diameter rotor with turbine blades. The etch that released the rotor, 15 μm wide by 300 μm deep, also created the hydrostatic gas journal bearing that provides radial support to the rotor. The second and fourth wafers included thrust bearings that provide axial support to the rotor. The outer wafers included channels to distribute flows and measure pressures throughout the device. Using pressurized nitrogen to drive the turbine, the rotor was spun to speeds as high as 1.4 million rpm (= 300 m s^{-1} tip speed). A description of the design and operation of this device is included in Ref. [8].

The yield of early devices was poor; in the build that contained the 300 m s^{-1} tip speed run, only 2 out of 12 devices achieved tip speeds greater than 100 m s^{-1}. The low yield resulted from inconsistent dimensional control across the wafer and an incomplete understanding of microbearing theory. A subsequent modeling effort greatly advanced knowledge of microbearing theory, explaining the relationship between bearing geometry and stability and showing how other elements within the device such as seals can couple into rotor stability [9–14]. In a more recent bearing demonstrator build, all tested devices (11 out of 12) achieved high speed [15, 16], with a peak speed of 1.7 million rpm (= $370\,\mathrm{m\,s^{-1}}$ tip speed).

The high speed devices discussed in the previous paragraph used hydrostatic gas thrust and journal bearings. In an effort to extend the range of bearing technologies available for these devices, Wong *et al.* [17] provided an initial demonstration of hydrodynamic gas thrust bearings.

5.3.1.2 Combustion

Combustion at the microscale is a challenge due to the limited volume available for the reaction to occur, leading to small fluid residence time within the combustor, which drives down combustor efficiency. Early work in this program demonstrated combustion of hydrogen in a silicon microcombustor [18]. However, hydrogen is not currently considered an attractive fuel for this engine due to its relatively low density, leading to difficulty in storage and transport.

MIT demonstrated gas-phase hydrocarbon (ethylene and propane) combustion, but at relatively low flow rates for a given combustor volume [19]. It was found that the ignition delay associated with hydrocarbon fuels makes a gas-phase reactor relatively large for microscale gas turbine engines. Spadacini *et al.* [20, 21] explored ways to limit combustor size. They showed that the introduction of a platinum catalyst into the combustor could extend the flow rate capacity by a factor of 3–4 for a given combustor volume. However, combustion efficiency was low, and modeling indicated that this was caused by the reaction being diffusion limited to the catalyst. Pressure drop in the catalytic combustor was low (~3%), providing some margin to increase efficiency by increasing catalyst density.

To achieve higher efficiency levels, a catalytically-enhanced gas phase combustor is planned, having a catalytic front-end followed by a gas-phase back-end. The catalytic reactor front-end (with even lower porosity) would raise gas temperature to the point where a gas-phase reactor would complete the reaction with minimal ignition delay. This scheme would also resolve temperature limitation issues associated with the catalyst and support structure material. An experiment to demonstrate this concept at the microscale is underway.

The MIT group is also demonstrating JP-8 micro-combustion. JP-8 is the fuel of choice of the U.S. Army for diesel and jet engines. It has low volatility and a volumetric energy density more than three times that of liquid hydrogen. The JP-8 microcombustor includes a vaporizer section that utilizes combustor heat loss into the structure to preheat and vaporize the incoming fuel, prior to the combustor.

5.3.1.3 Turbocharger

The turbocharger is a six-layer device, similar in design to the high-speed rotor demonstrator devices discussed above. However, the rotor in the turbocharger includes a turbine on one side and a compressor on the opposite side. The turbocharger is a second-generation device, serving as a test vehicle for the integration of bearings and turbomachinery. The turbocharger also includes a volume that

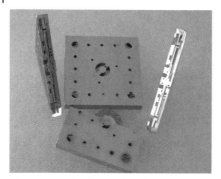

Figure 5.2 Turbocharger dies, formed from fusion bonding six silicon wafers. A full die (top center) is surrounded by sectioned dies that show internal flow paths. The bottom center die is sectioned, with a rotor reinserted. Dies are 23 mm × 23 mm × 2.9 mm.

could serve as a combustor. This device is designed to serve as a bridge from the earlier high-speed bearing demonstrators to a gas turbine engine.

The turbocharger rotor has a diameter of 8 mm (Figure 5.2). The rotor diameter was doubled from earlier generation devices to reduce viscous effects, improving turbomachinery performance. Savoulides *et al.* [22–24] spun a turbocharger rotor to 480 000 rpm (200 m s^{-1} tip speed). At this speed, the compressor achieved a pressure ratio of 1.21 with a flow rate of 0.13 g s^{-1}. To drive the compressor, the turbine was producing about 5 W of mechanical power. The spool efficiency (turbine and compressor combined) was 24% and increasing with speed. The turbocharger aerodynamic performance was shown to be consistent with design models. Experiments were also performed in which hydrogen was premixed with the turbine gas supply, allowing a flame to be lit in the combustor, upstream of the turbine, while the rotor was spinning. As expected, the rotor accelerated to a higher speed when the combustor was lit as higher enthalpy gas was now driving the turbine. Likewise, the rotor decelerated when the flame was extinguished.

5.3.1.4 Electric Generator

Several electric generator concepts have been evaluated for integration into the gas turbine. The electric generator would be mounted on a shroud on the compressor to locate it in the lowest temperature portion of the engine. Initially, an electro-quasistatic induction generator was pursued as it was thought to be more consistent with high-temperature operation than a magnetic generator. Electroquasistatic induction generators have been deemed impractical at the macroscale and had never been physically demonstrated. As part of this program, Steyn *et al.* [25, 26] achieved the first ever demonstration of this type of generator. However, it was found that the efficiency and power density of this generator concept were considerably lower than what could be achieved from magnetic machines.

A team from Georgia Tech, MIT and Clark Atlanta University investigated both magnetic induction and permanent magnet generators. A magnetic generator will have higher power density and higher efficiency than the electrostatic device. However, the high operating and fabrication process temperatures of the microscale gas turbine were originally a concern for a magnetic device since magnetic materials lose ferromagnetic strength when operated above their Curie temperature. Subsequent analysis showed that, given the thermodynamic need to design in features to thermally isolate the compressor from the engine hot section, the generator could be maintained below the Curie temperature during operation. While fusion bonding requires temperatures above the Curie temperature, experiments demonstrated that the magnets could survive exposure to these temperatures, although they would likely need to be remagnetized following exposure.

Tests on a magnetic induction generator showed that the power density was still too small for the gas turbine needs [27–30]. Permanent magnet machines were designed and fabricated using a hybrid microfabrication and assembly approach [31–33]. Initial tests were performed with microfabricated stators and conventionally machined rotors with SmCo magnets and a FeCoV (Hiperco 50) back iron inserted into them. The stators consisted of surface-wound electroplated copper coils (three-phase, eight pole) on ferromagnetic (NiFeMo) substrates. The rotor, supported on an air-driven spindle, demonstrated nearly 10 W of mechanical-to-electrical power conversion at a power density that was orders of magnitude better than the electric induction and magnetic induction machines. A MEMS turbine generator is currently being fabricated, integrating the high-speed bearing device process with the permanent magnet process.

5.3.1.5 Self-Sustaining Engine

MIT is now integrating the proven subcomponents into a MEMS gas turbine engine demonstrator. To reduce complexity and risk, the current version of the engine does not include an electric generator, although its design allows for later integration. The goal of this engine is to showcase the feasibility of MEMS gas turbines by demonstrating self-sustaining operation, a major milestone for all gas turbine engine programs.

Models developed and verified for the engine subsystems were combined into an overall engine system model. The goal was to find a design that would self-sustain with a minimum of changes in the fabrication process relative to the turbocharger device discussed earlier. Reducing heat transfer to the compressor was found to be critical, so the design includes a shaft between the compressor and turbine rotors. However, other performance improvements, such as silicon carbide deposition in the turbine to improve creep resistance [5], were deferred to future device generations.

The final design, shown in cross-section in Figure 5.3, contains 8 mm diameter compressor and turbine rotors connected by a 900 µm diameter by 660 µm long shaft, a substantially increased combustor volume relative to earlier devices, a design tip speed of 250–300 m s^{-1}, a combustor exit temperature of 1160 K, and a mass flow rate of $0.6\,g\,s^{-1}$. Earlier work demonstrated hydrogen microcombustors operating at temperatures in excess of 1160 K; operating the combustor at this relatively low temperature reduces creep concerns for the turbine, but also necessitates the increase in combustor volume. The tip speed of $250–300\,m\,s^{-1}$ translates to 600 000–700 000 rpm, which is in the range demonstrated by earlier devices. This design should allow the engine to break even, with the option of increasing combustor temperature to get more power at the cost of creep life. In order to generate significant excess mechanical power to drive an electric generator, later generation devices will need to operate at higher temperatures and higher tip speeds.

The cross-section shown in Figure 5.3 is not proportionally to scale to help clarify different components and to highlight several of the key features required for the gas turbine. The hatched region denotes the rotor. Air enters axially near the centerline, flows radially outward through a centrifugal compressor and diffuser, continues around the combustor to enhance thermal isolation, heats in the combustor, and then expands through nozzle guide vanes and a radial-inflow turbine rotor. Fuel is injected downstream of the diffuser. The rotor is supported axially by a pair of hydrostatic gas thrust bearings and radially by a pair of hydrostatic gas journal bearings. To reduce turning losses as the air enters the compressor, a molded curved inlet is inserted in the inlet region. An anodic bond process attaches the engine to a glass plate, thermally isolating the engine from packaging connections. The engine design includes ten fusion-bonded silicon wafers (shown on the right in Figure 5.3) formed from a set of 28 photolithography masks, several of which are used on multiple wafers. A first build cycle of this engine was completed with all process steps demonstrated, including the 10-wafer stack fusion bond. However, measurements taken during the build indicated that the journal bearing width was out of specification for high-speed operation. The MIT group is hoping to run additional build cycles of this engine to obtain testable specimens.

An additional advantage that the microscale gas turbine gains from scaling is a reduction in noise level relative to conventional scale gas turbines. Noise associated with the rotor blade passing frequency in a conventional engine is typically in the kHz frequency range. Since rotational frequency scales inversely with

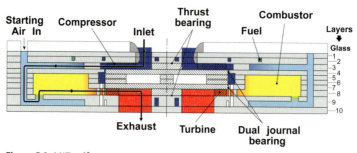

Figure 5.3 MIT self-sustaining engine cross-section, not to scale.

diameter, scaling the engine down by about a factor of 100 pushes rotor associated sound frequencies well above the audible range for humans. For instance, at a tip speed of $300 \, \text{m s}^{-1}$, the self-sustaining engine rotational frequency is about 12 kHz, and the blade passing frequency will be in excess of 60 kHz (depending on the actual number of blades).

Jet noise at the turbine exit is an additional concern. To reduce fluidic turning losses at the turbine exit, the engine is designed with a large exhaust nozzle, resulting in relatively low exit velocities. The low exit velocities should keep jet noise to a minimum.

One last issue to touch on is start-up. Because of the small scale and resulting small mass of the device, the rotor has relatively little rotational inertia and can thus accelerate quite rapidly to design speed, taking of the order of milliseconds. Thermal characteristics of the rotor and stator system define the start-up transient, which can last of the order of 10 s [22].

5.3.1.6 Other Gas Turbine Efforts

The desire to obtain the high power and energy density levels offered by gas turbines for portable power applications without the constraints of silicon microfabrication has led several groups to explore building small-scale gas turbines using traditional fabrication methods and materials. For instance, Isomura et al. designed a gas turbine generator fabricated from metal and ceramics using small-scale high precision milling [34, 35]. This gas turbine has 10 mm diameter turbine and compressor rotors connected by a 4 mm diameter shaft that is supported on gas bearings.

A group at the Katholieke Universiteit, Leuven developed a gas turbine generator that is EDM machined from stainless steel with an axial flow turbine and centrifugal compressor [36, 37]. This engine contains a 10 mm diameter axial-flow turbine and a 20 mm diameter centrifugal flow compressor. A shaft supported on conventional rolling contact bearings connects the turbomachinery rotors.

Kang et al. built a 12 mm diameter silicon nitride rotor using a mold shape deposition manufacturing (SDM) process [38]. The rotor contains a centrifugal compressor and radial inflow turbine, designed for use in a fist-sized gas turbine generator. They spun this device to 420 000 rpm and measured compressor flow characteristics that were consistent with models.

5.3.2
Rankine Engines

Fréchette and his group are developing microfabricated Rankine cycle-based electric generators [39, 40]. One of the key challenges for the microscale gas turbines described above is compressor performance, which is limited by small-scale effects such as viscous loss and heat transfer from the turbine. The Rankine cycle engine replaces the compressor with a liquid pump, which is less affected by small-scale physical effects and requires far less power to operate. A Rankine cycle engine offers the potential for a large gain in cycle efficiency with only a modest cost in

(a)   (b)

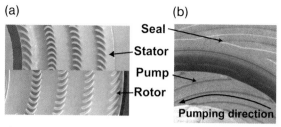

Figure 5.4 Turbine (left) and pump (right) for a Rankine-cycle based engine (reprint of Figure 5.2 from Ref. [39] by permission of the author and the Transducer Research Foundation).

power density, but with the additional complication of dealing with both liquid and gas phases of the working fluid. Figure 5.4 shows the multistage turbine and viscous pumping system developed by Lee *et al.* for a Rankine cycle application [39]. Cui and Brisson [41] also developed a model for a MEMS Rankine cycle engine that showed promising results.

5.3.3
Internal Combustion Engines

A team at U.C. Berkeley explored an Otto-cycle based MEMS rotary (Wankel) engine [42]. The engine design included a 2.4 mm diameter silicon rotor and an integrated magnetic electric generator. To reduce design complexity, rotors were fabricated separately from the stator, and the generator stator included discrete inserted components, thus requiring precision manual assembly. The Berkeley group also developed a scaled-up metal version of this engine to obtain higher power and demonstrate proof of concept.

Several development efforts are using small-scale conventional machining methods to fabricate internal combustion engines, two of which follow. Dahm *et al.* [43] demonstrated a swing engine that produces 20 W electric, which is being commercialized by Powerix Technologies. Aerodyne Research developed a 10 W two-cycle free piston engine [44]. Both of these engines fit easily in one's hand.

5.3.4
External Combustion Flexing Wall Heat Engine

The heat engines discussed to this point are scale reductions of macroscale machines. Richards and his group at Washington State University (WSU) are developing a microscale dynamic heat engine that is unlike any macroscale engine. The WSU engine is based on the expansion and compression of a saturated two-phase working fluid driven by an external heat source [45–47]. The fluid is enclosed in a cavity with flexible silicon walls. As the fluid undergoes a phase transition from liquid to gas, the pressure increase drives the silicon walls to flex out. A thermal switch cyclically contacts the cavity wall, inducing the phase transition. A piezoelectric film deposited on the silicon walls would produce electric power as it is flexed.

Figure 5.5 shows a cross-sectional sketch of the WSU engine concept. The flexing wall design of this engine is unsuitable for macroscale implementation. Rather, this engine gains its capabilities at the microscale. A single engine of this design has relatively low power (~milliwatts), so arrays of devices would be used to achieve the level of power desired for a given application. This engine concept can work over small or larger temperature differences. Thus this device could conceivably be used in an energy scavenging situation or with an external combustor. Modeling predicts a chemical-to-mechanical efficiency of about half that of Carnot efficiency [47].

Heat generated externally is transferred to the device by means of a piezoelectrically-driven thermal switch, as shown in Figure 5.5. The WSU group is exploring the use of a second thermal switch to actively cool the engine during the cycle stage in which it is desired to convert the gas back to a liquid. In a recent paper [45], the group presents results for a passively cooled engine operating at 5 Hz that generated 350 μW of mechanical power, while consuming only 7 μW to actuate the thermal switch. The heat source was held at a temperature of 60 °C. In a second test in which they incorporated active cooling, the engine was run to 100 Hz, generating 2.5 mW of mechanical power. For this second set of experiments, the WSU

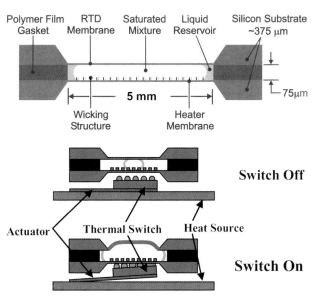

Figure 5.5 Cross-section of the WSU heat engine driven by a thermal switch (reprints of Figures 5.1 and 5.2 from Ref. [45] by permission of the author and the Institute of Physics Publishing).

group estimates that an optimized pair of thermal switches would have needed an actuation power of just 20 μW.

Acknowledgements

The MIT MicroEngine project was supported by the U.S. Army Research Laboratory, with early support from the Army Research Office and DARPA. The work on the MIT MicroEngine project included in this section represents the considerable effort of the MicroEngine team, the members of which are too numerous to list here. We thank them all for their assistance. The authors would also like to thank Luc Fréchette and Bob Richards for providing the information on their micro-heat engine projects that was included in this section.

References

1 Jacobson, S.A. (1998) Aerothermal challenges in the design of a microfabricated gas turbine engine, AIAA Paper 98-2545, 29th AIAA Fluid Dynamics Conference, Albuquerque, NM.

2 Waits, C.M., Morgan, B., Kastantin, M. and Ghodssi, R. (2005) Microfabrication of 3D silicon MEMS structures using gray-scale lithography and deep reactive ion etching. *Sensors and Actuators A (Physical)*, **119**, 245–53.

3 Ghodssi, R., Morgan, B. and Waits, C.M. (2005) Compensated aspect ratio dependent etching (CARDE) using gray-scale technology. *Microelectronic Engineering*, **77**, 85–94.

4 Chen, K.-S., Ayon, A. and Spearing, S.M. (2000) Controlling and testing the fracture strength of silicon on the mesoscale. *Journal of the American Ceramic Society*, **83**, 1476–84.

5 Moon, H.-S., Choi, D. and Spearing, S.M. (2004) Development of Si-SiC hybrid structures for elevated temperature micro-turbomachinery. *Journal of Microelectromechanical Systems*, **13**, 676–87.

6 Epstein, A.H. and Macro, Senturia, S.D. (1997) Power from micro machinery. *Science*, **276**, 1211.

7 Epstein, A.H. (2003) Millimeter-scale, MEMS gas turbine engines, Proceedings of ASME Turbo Expo, Paper GT-2003-38866, Atlanta, GA.

8 Fréchette, L.G., Jacobson, S.A., Breuer, K.S., Ehrich, F.F., Ghodssi, R., Khanna, R., Wong, C.W., Zhang, X., Schmidt, M.A. and Epstein, A.H. (2005) High-speed microfabricated silicon turbomachinery and fluid film bearings. *Journal of Microelectromechanical Systems*, **14**, 141–52.

9 Liu, L.X., Teo, C.J., Epstein, A.H. and Hydrostatic, Z.S. (2005) Hydrostatic gas journal bearings for micro-turbomachinery. *Journal of Vibration and Acoustics*, **127**, 157–64.

10 Liu, L.X. and Spakovszky, Z.S. (2007) Effects of bearing stiffness anisotropy on hydrostatic micro gas journal bearing dynamic behavior. *Journal of Engineering for Gas Turbines and Power*, **129**, 177–84.

11 Liu, L.X. (2005) Theory for hydrostatic gas journal bearings for micro-electro-mechanical systems, MIT PhD Thesis.

12 Teo, C.J., Spakovszky, Z.S. and Jacobson, S.A. (2006) Unsteady flow and dynamic behavior of ultra-short lomakin gas bearings. Proceedings of STLE/ASME International Joint Tribology Conference, IJTC 2006.

13 Teo, C.J. and Spakovszky, Z.S. (2006) Modeling and experimental investigation of micro-hydrostatic gas thrust bearings for micro-turbomachines. *Journal of Turbomachinery*, **128**, 597–605.

14 Teo, C.J. and Spakovszky, Z.S. (2006) Analysis of tilting effects and geometric nonuniformities in micro-hydrostatic gas thrust bearings. *Journal of Turbomachinery*, **128**, 606–15.

15 Teo, C.J., Liu, L.X., Li, H.Q., Ho, L.C., Jacobson, S.A., Ehrich, F.F., Epstein, A.H. and Spakovszky, Z.S. (2006) High-speed operation of a gas-bearing supported MEMS air turbine. Proceedings of STLE/ASME International Joint Tribology Conference, IJTC 2006.

16 Teo, C.J. (2006) MEMS turbomachinery rotordynamics: modeling, design and testing, MIT PhD Thesis.

17 Wong, C.W., Zhang, X., Jacobson, S.A. and Epstein, A.H. (2004) A self-acting gas thrust bearing for high-speed microrotors. *Journal of Microelectromechanical Systems*, **13**, 158–64.

18 Mehra, A., Zhang, X., Ayon, A.A., Waitz, I.A., Schmidt, M.A. and Spadaccini, C.M. (2000) A six-wafer combustion system for a silicon micro gas turbine engine. *Journal of Microelectromechanical Systems*, **9**, 517–27.

19 Spadaccini, C.M., Mehra, A., Lee, J., Zhang, X., Lukachko, S. and Waitz, I.A. (2003) High power density silicon combustion systems for micro gas turbine engines. *Journal of Engineering for Gas Turbines and Power*, **125**, 709–19.

20 Spadaccini, C.M., Peck, J. and Waitz, I.A. (2007) Catalytic combustion systems for microscale gas turbine engines. *Journal of Engineering for Gas Turbines and Power*, **129**, 49–60.

21 Spadaccini, C.M., Zhang, X., Cadou, C.P., Miki, N. and Waitz, I.A. (2003) Preliminary development of a hydrocarbon-fueled catalytic micro-combustor. *Sensors and Actuators A (Physical)*, **A103**, 219–24.

22 Savoulides, N. (2004) Development of a MEMS turbocharger and gas turbine engine, MIT PhD Thesis.

23 Savoulides, N., Jacobson, S.A., Li, H., Teo, C.J. and Epstein, A.H. (2006) Performance of a high-speed microscale turbocharger, Proceedings of the 6th International Workshop on Micro and Nanotechnology for Power Generation and Energy Conversion Applications (PowerMEMS 2006).

24 Savoulides, N., Jacobson, S.A., Li, H., Ho, L., Khanna, R., Teo, C-J., Protz, J.M., Wang, L., Ward, D., Schmidt, M.A. and Epstein, A.H. (2008) Fabrication and testing of a high-speed microscale turbocharger. *Journal of Microelectromechanical Systems*, **17**, 1270–82.

25 Steyn, J.L. (2005) A microfabricated ElectroQuasiStatic induction turbine-generator, MIT PhD Thesis.

26 Steyn, J.L., Kendig, S.H., Khanna, R., Lyszczarz, T.M., Umans, S.D., Yoon, J.U., Lang, H.J. and Livermore, C. (2005) Generating electric power with a MEMS electroquasistatic induction turbine-generator, Proceedings of the 18th IEEE International Conference on Micro Electro Mechanical Systems (MEMS). pp. 614–17.

27 Arnold, D.P., Das, S., Cros, F., Zana, I., Allen, M.G. and Lang, J.H. (2006) Magnetic induction machines integrated into bulk-micromachined silicon. *Journal of Microelectromechanical Systems*, **15**, 406–14.

28 Koser, H. and Lang, J.H. (2006) Magnetic induction micromachine – Part I: design and analysis. *Journal of Microelectromechanical Systems*, **15**, 415–26.

29 Cros, F., Koser, H., Allen, M.G. and Lang, J.H. (2006) Magnetic inducstion micromachine-part II: fabrication and testing. *Journal of Microelectromechanical Systems*, **15**, 427–39.

30 Koser, H. and Lang, J.H. (2006) Magnetic induction micromachine-part III: eddy currents and nonlinear effects. *Journal of Microelectromechanical Systems*, **15**, 440–56.

31 Arnold, D.P., Herrault, F., Zana, I., Galle, P., Park, J.-W., Das, S., Lang, J.H. and Allen, M.G. (2006) Design optimization of an 8 W, microscale, axial-flux, permanent-magnet generator. *Journal of Micromechanics and Microengineering*, **16**, S290–6.

32 Das, S., Arnold, D.P., Zana, I., Park, J.-W., Allen, M.G. and Lang, J.H. (2006) Microfabricated high-speed axial-flux multiwatt permanent-magnet generators – Part I: modeling. *Journal of Microelectromechanical Systems*, **15**, 1330–50.

33 Arnold, D.P., Das, S., Park, J.-W., Zana, I., Lang, J.H. and Allen, M.G. (2006) Microfabricated high-speed axial-flux multiwatt permanent-magnet Generators-part II: design, fabrication, and testing. *Journal of Microelectromechanical Systems*, **15**, 1351–63.

34 Isomura, K., Murayama, M., Teramoto, S., Hikichi, K., Endo, Y., Togo, S. and Tanaka, S. (2006) Experimental verification of the feasibility of a 100 W class micro-scale gas turbine at an impeller diameter of 10 mm. *Journal of Micromechanics and Microengineering*, **16**, S254–61.

35 Isomura, K., Tanaka, S., Togo, S., Kanebako, H., Murayama, M., Saji, N., Sato, F. and Esashi, M. (2004) Development of micromachine gas turbine for portable power generation. *JSME International Journal Series B*, **47**, 459–64.

36 Peirs, J., Reynaerts, D. and Verplaetsen, F. (2004) A microturbine for electric power generation. *Sensors and Actuators A: Physical*, **113**, 86–93.

37 Peirs, J., Reynaerts, D. and Verplaetsen, F. (2003) Development of an axial microturbine for a portable gas turbine generator. *Journal of Micromechanics and Microengineering*, **13**, S190–5.

38 Kang, S., Matsunaga, M., Johnston, J.P., Tsuru, H., Arima, T. and Prinz, F.B. (2003) Micro-scale radial-flow compressor impeller made of silicon nitride – manufacturing and performance. *Proceedings of the ASME Turbo Expo*, **3**, 779–88.

39 Lee, C., Liamini, M. and Fréchette, L. (2006) Design, fabrication, and characterization of a microturbopump for a Rankine cycle micro power generator. Proceedings of the Solid State Sensors, Actuators and Microsystems Workshop, Hilton Head Island, SC.

40 Fréchette, L.G., Lee, C., Arslan, S. and Liu, Y.-C. (2003) Design of a microfabricated Rankine cycle steam turbine for power generation. *Proceedings of the ASME International Mechanical Engineering Congress, Micro-Electromechanical Systems Division*, **5**, 335–44.

41 Cui, L. and Brisson, J.G. (2005) Modeling of MEMS-type Rankine cycle machines. *Journal of Engineering for Gas Turbines and Power*, **127**, 683–92.

42 Walther, D.C. and Pisano, A.P. (2003) MEMS rotary engine power system: project overview and recent research results. Proc. 4th Intl Symposium on MEMS and Nanotechnology, pp. 227–34.

43 Dahm, W.J.A., Ni, J., Mijit, K. Mayor, R., Qiao, G., Benjamin, A., Gu, Y., Lei, Y. and Papke, M. (2002) Internal Combustion Swing Engine (MICSE) for portable power generation systems. 40th AIAA Aerospace Sciences Meeting, AIAA Paper 2002-0722.

44 Annen, K.D., Stickler, D.B. and Woodroffe, J. (2002) Miniature Internal Combustion Engine (MICE) for portable electric power. 23rd Army Science Conference.

45 Cho, J.H., Weiss, L.W., Richards, C.D., Bahr, D.F. and Richards, R.F. (2007) Power production by a dynamic micro heat engine with an integrated thermal switch. *Journal of Micromechanics and Microengineering*, **17**, S217–23.

46 Weiss, L.W., Cho, J.H., McNeil, K.E., Richards, C.D., Bahr, D.F. and Richards, R.F. (2006) Characterization of a dynamic micro heat engine with integrated thermal switch. *Journal of Micromechanics and Microengineering*, **16**, S262–9.

47 Whalen, S., Thompson, M., Bahr, D., Richards, C. and Richards, R. (2003) Design, fabrication and testing of the P3 micro heat engine. *Sensors and Actuators A: Physical*, **104**, 290–8.

6
Thermophotovoltaics

Ole Nielsen

6.1
Overview

Thermophotovoltaic (TPV) generation of electricity is based on a heated emitter radiating photons that are converted to electricity by photocells. The concept was proposed as early as 1956 by H.H. Kolm at the MIT Lincoln Laboratories [1]. It did not, however, gain wide popularity as a research area until the late 1980s, when more suitable low-bandgap photocell materials became available. In recent years much of the TPV related research has focused on optimizing photocell and emitter materials and structures, as well as filters [2–4] and back reflectors [5] for photon recycling. Complete systems have been assembled for applications ranging from waste heat harvesting in furnaces [6] to portable power generators [7–9]. System efficiencies as high as 12.3%, have been achieved [6].

The TPV principle is analogous to solar cell technology, the main difference being that a locally heated emitter replaces the sun as a source of radiation for the photocell. Since, for a TPV system, the photocells can be placed in close proximity to the emitter, the radiated power density can be 150–1500 times higher than in solar cell systems, despite the much lower heat source temperature. However, because the emitter temperature is inferior to that of the surface of the Sun, the photons emitted in a TPV system have lower energy, hence different photocell materials must be used to convert the radiated energy to useful electrical energy.

The advantages of TPV systems over other portable power generation schemes are manifold. First and foremost, it is a mechanically passive technology, which makes it simpler to construct and less subject to wear than engines and turbines. It is also practically noiseless. Second, there is no physical contact between the hot and cold zones of the device. As a consequence, TPV systems do not exhibit problems associated with large thermal stresses. In contrast, thermoelectric (TE) devices rely on thermopiles physically connecting the heat source to the thermal sink [10]. Additionally, this feature allows the photocells on the one hand and the combustor/emitter on the other hand, to be fabricated separately and integrated later. This greatly simplifies fabrication, since photocells can be complex material structures that could be very difficult to integrate into another fabrication process.

Microfabricated Power Generation Devices. Edited by Alexander Mitsos and Paul I. Barton
Copyright © 2009 WILEY-VCH Verlag GmbH & Co. KGaA, Weinheim
ISBN: 978-3-527-32081-3

Finally, the chemistry associated with the heating of the emitter is usually limited to a simple catalytic combustion. This tends to make design and construction of TPV systems simpler than fuel cells, where complicated chemistries and materials also limit the lifetime of the device, especially when subjected to thermal cycling (on–off switching).

Challenges of TPV include the need to maintain a high emitter temperature (typically higher than 700 °C) and a relatively low photocell temperature (typically less than 50 °C), relatively low efficiencies and very limited commercial availability of photocells within the desirable range of bandgaps.

Portable TPV power generators are not yet commercially available and those that have been developed are relatively large. Scaling down TPV generators opens up a new array of potential applications. There are no fundamental problems caused by scaling, since the optics used at this size are identical to those at large scales. In fact, since the surface-to-volume ratio increases with decreasing dimensions, and TPV power output is proportional to emitter/photocell surface area, nominally the volumetric power density increases when scaling down.

6.2
Thermophotovoltaic System Components

A conceptual schematic of a TPV system is shown in Figure 6.1. It shows the main system components and heat flow paths in a portable TPV generator. Green arrows signify useful power input and output, orange arrows signify optical heat flow, and the red arrows detail the various undesirable heat losses. The efficiency of a TPV system is defined in different ways, depending on the application. In waste heat TPV systems [6] and in photocell tests (not complete systems), it is defined as the ratio of electrical power produced to the power radiated from the emitter. This does not include other heat loss paths from the emitter structure. In portable generators, efficiency is typically defined as the ratio of the electrical power produced to the power input in the form of heat from the combustion. Ultimately,

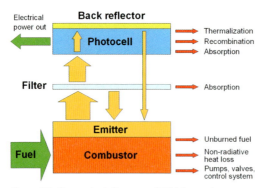

Figure 6.1 Conceptual diagram of TPV thermal management.

power spent on auxiliary tasks such as controlling gas flow, pumping air and photocell cooling, must be included for the efficiency to be a good measure of the performance of a particular portable power source. These are often not included in the reported numbers and it is important to distinguish between different definitions of efficiency before comparing TPV systems.

6.2.1
Photocell Materials

In a photocell (or photo diode) impinging photons with energies larger than or equal to the bandgap of the material contribute to the photo current, which provides the electrical power output. TPV emitters mostly radiate infrared light, so photocell materials with low bandgaps are required to utilize a considerable portion of the radiated spectrum. Silicon photocells, although well suited for the mostly visible solar spectrum, are not generally useful for TPV because the bandgap of silicon (1.1 eV) is too large. Suitable materials for TPV include GaSb, InAs, and various ternary and quaternary semiconductors with bandgaps around 0.7 eV and lower [5, 8, 11, 12].

   Photocells are transparent to photons with energy lower than the bandgap. Each photon with energy higher than the bandgap energizes an electron to produce one electron–hole pair that contributes to the electrical current. Any energy above the bandgap results in additional thermalization of the carriers and heating of the cell, without providing additional electrical power. It would therefore be ideal for the energy conversion efficiency if all the incident photons were located slightly above the bandgap in the energy spectrum. However, since radiation at any given wavelength is limited by the black body spectrum at a certain temperature, limiting the emitted spectrum to a narrow peak will result in very low power density. In the context of photocells, power density is given by area ($W m^{-2}$), since the power output is proportional to the photocell surface area upon which the radiation is incident.

6.2.2
Emitters

The emitter, the source of radiation in the TPV system, is typically heated by the combustion of fuels. Since radiation is largely a surface phenomenon, it is important to ensure that the surface of the emitter not only exhibits a high temperature but that it also has appropriate material and structural characteristics. Emitter types are typically divided into three categories: broadband, matched and selective [11, 13].

   Broadband emitters radiate over the full blackbody spectrum, although at an emissivity lower than 1. They are also known as gray body emitters. Their advantage is that they have the overall highest power density (since they radiate in the entire black body spectrum) but this is to the detriment of the efficiency of the TPV system. Photons of energy lower than the bandgap remain unused, and at much higher energies, the photons cause considerable thermalization in the photocells. To improve the efficiency, filters on the front and mirrors on the back of the photocells can be used to recycle high and low energy photons back to the emitter. Many materials can be approximated as gray body emitters. A gray body radiates a spectrum identical in shape to the black body spectrum for a given temperature, scaled down by a factor ε known as the emissivity (a number between 0 and 1, 1 being an ideal black body). When a gray body is illuminated, it absorbs a certain fraction of the incident light. This is called the absorptance, denoted by α, and is equal in value to the emissivity. The rest of the light is either reflected or transmitted.

Matched emitters shape the emitted spectrum without the use of filters. This can be achieved by using materials with special optical properties. For instance, the radiative spectra of certain refractory metals contain very few photons below the bandgap of GaSb [6], resulting in a matched emitter/photocell pair. Photonic crystal structures have also been built on the surface of emitters to control the spectrum [3, 13, 14]. When constructed correctly, matched emitters tend to provide a good trade-off between efficiency and power density.

Lastly, selective emitters radiate in a very narrow band just above the bandgap of the photocell material. Rare earth oxides are typically chosen as emitter materials for this purpose, since they have a peculiar radiation spectrum with a single narrow peak. In some cases, this peak matches well with the bandgap of a certain photocell material. An example of such a pair is ytterbia (emitter) and silicon (photocell) [11]. Selective emitters usually provide high efficiencies but low power densities.

In general, the performance (efficiency and power density) of TPV systems improves with increased temperature, for two reasons. First, the emitted power is proportional to the fourth power of temperature. Second, except with selective emitters, the peak of the spectrum shifts upwards in energy, allowing a higher percentage of the photons to be turned into electrical power. Typically, it is desirable for the peak of the emitter spectrum to lie somewhat above the bandgap of the photocell material. The photocell material must therefore be chosen according to the emitter temperature, and the temperature of the photocell itself must be kept low (typically not far above room temperature). The recombination rate in the photocell substrate increases with temperature, negatively impacting the efficiency.

6.3
Thermal Management in TPV Systems

The temperature required for an emitter in a TPV system normally exceeds that of any of the other energy conversion technologies, because it relies on radiation, which is proportional to the fourth power of temperature. To maintain this high temperature in a given zone, the emitter must be extremely well thermally isolated from the surroundings and, as in most cases the heat will be provided by a com-

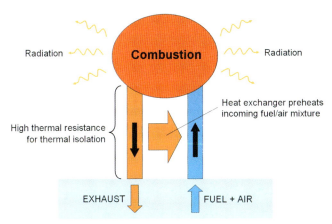

Figure 6.2 General TPV combustor–emitter thermal requirements.

bustion reaction, gas flows must be allowed to and from it. For the isolation to be effective, heat exchange from the exhaust gas to the incoming gas is a necessity. The only desired heat flow path in a TPV system is radiation, which is therefore enhanced rather than suppressed (as opposed to other types of generators). This is illustrated in Figure 6.2. The pathways for heat loss in any generator system are the following: radiation, solid conduction, fuel gas forced convection and ambient gas conduction and natural convection. Radiation, if undesired, can be reduced by reflective coatings on the hot zone surface or by surrounding it with reflective surfaces. The emissivity and absorptance of a material are the same. Therefore a highly reflective coating also has a very low emissivity. In TPV systems highly absorbing (or dark) emitter surfaces are therefore desired. Solid conduction is usually reduced by applying thick layers of insulation, that is, materials with very low thermal conductivity but, since this is impractical in most microsystems, long and narrow structures with a high thermal resistance may be used instead. Heat losses by fuel gas forced convection in the system can be reduced by using effective heat exchangers between the incoming and outgoing gas flows. Finally, ambient gas conduction and natural convection can be reduced by surrounding the hot zone with a less conducting gas than air, or can be eliminated by vacuum packaging. In both cases the system must be *fully closed*, meaning that no open flames can be used.

6.4
Scaling of TPV Systems

In the context of TPV technology, scaling down existing technology presents both challenges and advantages due to the particular ways that the physics change with decreasing dimensions. The important aspects to consider are in the optical, thermal and fluid mechanics domains. Since the technology does not require moving parts in critical areas, the purely mechanical aspects are not generally of great importance. Particular fabrication technologies may, however, impose more constraints on the scaling.

6.4.1
Optics

The optics involved in TPV systems are typically limited to radiative heat transfer between two or more surfaces. The surfaces may be complicated systems in themselves, with coatings, filters or photonic crystals to tailor the spectrum emitted or absorbed. However, the structures of these are determined by the spectrum (wavelength), which is independent of the overall size of the system.

In any TPV system it is assumed that a view factor close to 1 can be achieved between the emitter and the photocells. This advantage is gained by "scaling down" the emitter from the Sun (solar cells) to a local, containable heat source. Scaling down of TPV systems does not offer any further improvements in that area. The same holds true for the optical cavity that is inevitably created in a high view factor, three-dimensional system. At very small scales it may actually be more difficult to achieve a high view factor due to fabrication constraints.

Optically, other than the particular shape of a system, the scaling problem then reduces to shrinking the gap between the emitter and the photocell. Other than changing the optical cavity, shrinking this gap has no effect on the radiative transfer until the gap size approaches the wavelength of the light emitted. It has been shown theoretically [15] and experimentally [16] that, in this case, a coupling of the evanescent field of the emitter to the photocell surface can greatly increase the radiative transfer. While this may have very interesting applications it poses certain difficulties in controlling this extremely small gap.

6.4.2
Heat Transfer

The most profound effect of scaling on the heat transfer comes from the change in surface-to-volume ratio. Heat transfer, be it radiative, conductive or convective, is overall proportional to surface area. Therefore, the heat loss relative to system size and weight is directly proportional to the surface-to-volume ratio of the device. Regardless of the shape of the system, scaling down the dimensions, without otherwise changing the geometry, leads to a linear *increase* in this ratio.

This has two significant implications. First, the heat losses from the system, relative to its size, will increase by the inverse of the system scaling factor, which in turn lowers the efficiency of the generator. It is therefore crucial to improve the thermal isolation of the reactor or hot zone. As discussed, this is typically done, in larger systems, with thick layers of insulating material. But with scaling down, for the very same reasons, this insulation becomes much less effective. Keeping the insulating material at the same thickness as in the larger system would ensure good thermal isolation, but not only might it be practically impossible to design, it would also increase the system size and weight manifold. Other methods, such as high aspect ratio structures [17] may therefore be used instead.

The second implication of increased surface-to-volume ratio and the associated increase in heat transfer presents an advantage to TPV systems. Since TPV power output is directly proportional to the surface area (of the photocells and the emitter), increasing the surface-to-volume ratio leads to an increased volumetric power density S, as shown in Equation 6.1.

$$S = s\frac{A}{V} \tag{6.1}$$

where s is the power per unit area of the emitter–photocell pair, which is dependent on the emitted spectrum, the photocell material and structure, and the view factor, but not on scaling. While this increase in volumetric power density is an added advantage of smaller systems, it also suggests that larger systems could benefit from being replaced by clusters of smaller ones, thus achieving an overall higher power density. Again, this depends further on the limitations of the fabrication technology.

Another advantage to smaller systems is that vacuum packaging is more feasible. The purpose of vacuum packaging is to remove the air conduction and natural convection heat loss path between the hot and cold parts of the generator. What constitutes an effective vacuum level for a particular system depends on the gas and the size of the cavity between the cold and hot zones. To achieve an effective reduction in gas conduction and natural convection, the mean free path must be longer than the largest cavity dimension D. Equation 6.2 gives the mean free path of an ideal gas, where λ is the mean free path, k_B is the Boltzmann constant, T is temperature, P is pressure and d is the average diameter of the molecules constituting the gas.

$$\lambda = \frac{k_B T}{\sqrt{2}\pi d^2 P} \tag{6.2}$$

Combining this equation with the condition $\lambda > D$ then imposes the following condition on the pressure:

$$P < \frac{k_B T}{\sqrt{2}\pi d^2 D} \tag{6.3}$$

As a consequence of smaller cavities not requiring as low a pressure level, portable generators can benefit from vacuum packaging, whereas most large systems have to resort to other means, such as using an ambient gas with lower heat conductivity.

Coupling of the evanescent field through sub-wavelength gap size has the potential advantage of greatly increased radiative (useful) heat transfer for TPV systems (theoretically by up to a factor of about 10). However, the challenges associated with this are many. First and foremost it is essential to employ vacuum packaging, since the thermal resistance of any gas in between the two surfaces is minimal and would result in very large gas conduction heat losses. Second, controlling the gap size without touching the surfaces directly (which would be catastrophic to the photocells) is complicated. This has so far been achieved by using very thin silicon dioxide spacers but, despite being very small, the diameter is greater than the thickness, which leads to a very low thermal resistance. No solution for this challenge has yet been published. Additionally, if physical contact through spacers is necessary, one of the big advantages of TPV (physical separation of the hot and cold zones) is removed, introducing challenges with respect to thermal expansion mismatch. However, if this hurdle could be overcome, evanescent coupling could provide a path to higher power and higher efficiency TPV systems.

6.4.3
Conclusion on Scaling

Treating portable power generators as a case of scaling down already existing technology is, at best, a good starting point or partial solution for TPV-based systems. While the approach can be used to get important answers and guide design to some extent, it is important to realize that key differences exist that require new thinking in design and can be taken advantage of only by deviating from traditional design rules. Furthermore, at sufficiently small scales, fabrication techniques impose very different constraints on geometrical shapes and layout than with large scale machining. This limits how directly scaling can be applied.

6.4.4
Examples of TPV Micro-Generator Systems

The general field of TPV has seen surges and declines of popularity and funding since its emergence in the late 1980s. Much of the research has been focused on developing low bandgap photodiodes capable of converting radiated energy from low-temperature (relative to the Sun) heat sources. As this crucial and enabling technology has progressed, attention has also been given to the potential for tailoring emitters for use with particular photocell materials. Since the emitter and photocell (along with filters) constitute the actual energy conversion component, they are naturally attracting the vast majority of the effort in this still emerging field. A few projects have built (or attempted to build) semi-complete systems that include the gas combustion heat source and, of these, only two published results seem to fall into the category of microsystems, dealing with the particular challenges discussed in the previous sections. These two will be discussed here to highlight what has been attempted so far and which areas the two projects have concentrated on optimizing.

6.4.4.1 A Prototype Microthermophotovoltaic Power Generator
This project by W. M. Yang et al. [18] has developed an open, tube-shaped micro-combustor that burns hydrogen (Figure 6.3). The particular geometry of the tube allows a very uniform temperature to develop during combustion, something that is desirable for high conversion efficiencies in the photocells. The tube is covered

Figure 6.3 Tubular microcombustor running hydrogen combustion. Reprinted with permission from [18]. Copyright 2004, American Institute of Physics.

with a layer of SiC (silicon carbide), which has a high emissivity (about 0.9). The project has clearly steered away from the use of MEMS fabrication technology to avoid the material and geometric limitations that come with it. More conventional processes have been used to create the tubular structure, probably at lower cost and more simplicity. The tube emitter is surrounded by a hexagonal arrangement of GaSb photocells (provided by JXCrystals Inc.) with outward facing heat sinks, achieving a high view factor (Figure 6.4). The uniform and high temperature of the emitter, along with the cylindrical arrangement with a high view factor from emitter to photocells, constitute the main optimization of this set-up.

The main apparent limitations of this approach are twofold: the open combustor and the use of hydrogen as the fuel. The combustor being open at one end to hot

Figure 6.4 Prototype micro-TPV power generator photocell array with cooling fins. Reprinted with permission from [18]. Copyright 2004, American Institute of Physics.

exhaust has several implications. First, the heat of the exhaust cannot be easily recycled to preheat the inlet gases. Second, vacuum packaging will be hard to achieve without making direct physical contact between the coolest and hottest zones of the system. Both greatly inhibit an effective thermal isolation of the hot zone. Heat conduction down the tube is also likely to be considerable, since very thin tubes walls are hard to achieve without resorting to microfabrication. Presumably a low thermal efficiency is the reason why hydrogen was used as fuel. It burns more easily and hotter, and autothermal combustion of hydrocarbon fuels such as propane or butane was likely not possible. Hydrogen is hard to compress and is not a viable fuel for portable generator systems unless it is carried in another form (e.g. ammonia) and converted *in situ*, requiring an additional, complex conversion system.

A total of over 1 W of electrical power output was achieved with an emitter temperature of 1052 °C, and the highest system efficiency reported was 0.66%. While continuing to refine the photocell structure and the emitter materials, it would be crucial to address the thermal isolation of this system. If fundamental improvements are made in that respect then this will be a promising concept.

6.4.4.2 A Thermophotovoltaic Micro-Generator for Portable Power Applications

This project by O. M. Nielsen *et al.* [17] has taken a different approach to the TPV system, using the suspended microreactor (SµRE), shown in Figure 6.5. Emphasis is first and foremost on thermal isolation between hot and cold zones. By using MEMS fabrication technology, a *closed* catalytic combustor is suspended from the surrounding substrate by long, high aspect ratio tubes made of silicon nitride, which has low heat conductivity. These tubes allow gases to enter and exit the combustor while recovering exhaust heat with heat exchangers connecting adjacent tubes. Being physically separated from the chemical reaction, the cavity around the reactor lends itself well to vacuum packaging. The excellent thermal isolation allows this device to burn propane and air autothermally, even without being in a vacuum. GaSb photocells (also from JXCrystals Inc.) were used to convert radiation to electrical power.

The SµRE goes a long way towards solving the thermal isolation problem, not only of TPV systems, but high-temperature microsystems in general. Its useful-

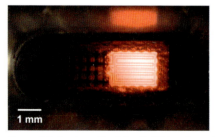

Figure 6.5 A schematic of the SµRE together with a picture of it heated to about 900 °C, substrate maintaining about 50 °C.

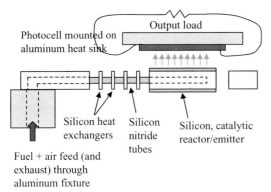

Figure 6.6 SμRE TPV power generation test set-up schematic.

ness for TPV is clear, but it does have several limitations that need to be addressed to make it successful in that regard. First, photocells can only be placed above and below the combustion chamber, leaving radiation from the sides to be absorbed by the surrounding substrate and thus lowering the view factor and the overall system efficiency. There are also material constraints linked to the fabrication processes used. The main structural material, silicon, has limited temperature capability and becomes soft and may undergo plastic deformation above 1000 °C. This makes the use of GaSb less desirable, and other photocell materials are not readily available. Additionally, the emissivity of this silicon nitride covered reactor is close to 0.7, which is considerably lower than that of SiC.

A total power output of 1 mW was achieved, which corresponds to a power density of 32 mW m^{-2} and efficiency of 0.08%. A schematic of the test set-up is shown in Figure 6.6. The test conditions were with only one photocell (above), in air, with the emitter at 770 °C, and it was estimated that with photocells on both sides, in vacuum and at 1000 °C, this would increase the power density and efficiency to more than 250 mW m^{-2} and 2.4%, respectively. Improvements in particular in the view factor and also photocell materials and emissivity, could make this a promising TPV system concept.

6.5
Conclusion

TPV as an energy conversion technology for micro-generator systems may be a viable choice for certain applications. The low efficiency prospects limit its potential use in systems where large heat output relative to that of the load device is not of great concern (such as external power supplies). The main attraction of TPV is material simplicity beyond that of the photocells themselves. The physical separation of hot and cold areas, along with the simple chemical reactions involved and no moving parts, allows a relatively simple fabrication process and integration as compared with other conversion methods. No portable TPV generators are as yet commercially available, and although some interesting attempts have been made at assembling such systems at the microscale, the research field is currently getting less attention than, for example, fuel cells.

Acknowledgments

This work was supported by the DoD Multidisciplinary University Research Initiative (MURI) program administered by the Army Research Office under Grant DAAD19-01-1-0566. The author would like to thank his thesis advisors Martin A. Schmidt and Klavs F. Jensen for their invaluable support.

References

1 Kolm, H.H. (1956) Quarterly progress report, solid state research, group 35, Technical report, MIT Lincoln Laboratory, Lexington, MA, May.

2 Brown, E.J. (2003) The status of thermophotovoltaic energy conversion technology at Lockheed Martin Corp, Space Technology and Applications International Forum, Albuqureque, NM, USA, February.

3 Celanovic, I. (2006) Thermophotovoltaics: shaping the flow of thermal radiation, PhD dissertation, Massachusetts Institute of Technology, Department of Chemical Engineering.

4 Fraas, L. (2002) Thermophotovoltaics for combined heat and power using low NOx gas fired radiant tube burners, Thermophotovoltaic Generation of Electricity, 5th Conference, number 653 in AIP Conference Proceedings, Rome, Italy, September, pp. 61–70.

5 Wang, C.A. (2002) Lattice-matched GaInAsSb/AlGaAsSb/GaSb materials for thermophotovoltaic devices, Thermophotovoltaic Generation of Electricity, 5th Conference, number 653 in AIP Conference Proceedings, Rome, Italy, September, pp. 324–34.

6 Fraas, L.M., Avery, J.E. and Huang, H.X. (2002) Thermophotovoltaics: heat and electric power from low bandgap solar cells around gas fired radiant tube burners, 29th IEEE PVSC Conference, New Orleans, USA, May.

7 Durisch, W. (2002) Small thermophotovoltaic prototype systems, Thermophotovoltaic Generation of Electricity, 5th Conference, number 653 in AIP Conference Proceedings, Rome, Italy, September, pp. 71–8.

8 Horne, E.E. (2002) 500 watt diesel fueled TPV portable power supply, Thermophotovoltaic Generation of Electricity, 5th Conference, number 653 in AIP Conference Proceedings, Rome, Italy, September, pp. 91–100.

9 Nelson, R.E. (2002) TPV and state-of-art development, Thermophotovoltaic Generation of Electricity, 5th Conference, number 653 in AIP Conf. Proc., Rome, Italy, September, pp. 3–17.

10 Schaevitz, S.B., Franz, A.J., Jensen, K.F. and Schmidt, M.A. (2001) A combustion-based MEMS thermoelectric power generator, Transducers '01, Munich, Germany, June.

11 Coutts, T.J. (1999) A review of progress in thermophotovoltaic generation of electricity. *Renewable and Sustainable Energy Reviews*, 3, 77–184.

12 Wernsman, B. (2004) Greater than 20 thermophotovoltaic radiator/module system using reflective spectral control. *IEEE Transactions on Electron Devices*, 51 (3), 512–5.

13 Gombert, A. (2002) An overview of TPV emitter technologies, Thermophotovoltaic Generation of Electricity, 5th Conference, number 653 in AIP Conference Proceedings, Rome, Italy, September, pp. 123–31.

14 Lin, S.Y. (1998) A three-dimensional photonic crystal operating at infrared wavelengths. *Nature*, **394**, 251–3.

15 Pan, J.L., Choy, H.K.H. and Fonstad, C. G. (2000) Very large radiative transfer over small distances from a black body for thermophotovoltaic applications. *IEEE Transactions on Electron Devices*, **47** (1), 241–49.

16 DiMatteo, R.S. (2002) Micron-gap thermophotovoltaics (MTPV), Thermophotovoltaic Generation of Electricity, 5th Conference, number 653 in AIP Conference Proceedings, Rome, Italy, September, pp. 232–40.

17 Nielsen, O.M., Arana, L.R., Baertsch, C.D., Jensen, K.F. and Schmidt, M.A. (2003) A thermophotovoltaic micro-generator for portable power applications, Transducers '03. Boston, USA, June.

18 Yang, W. (2004) A prototype microthermophotovoltaic power generator. *Applied Physics Letters*, **84** (19), 3864–6.

7
Thermal Management and System Integration

Benjamin A. Wilhite

7.1
Introduction

Heat management of power generation devices is critical to maximizing thermal efficiencies. Successful heat management requires consideration of the overall portable power system, identification of integration schemes for each component and selection of appropriate materials and packaging. As will be discussed in this chapter, heat management strategies must often incorporate each of these aspects in tandem to achieve maximum thermal efficiencies.

At the heart of any portable power generation device is the fuel from which the power will be derived, and *fuel selection often dictates the overall design of the power generation device*. During the course of the past 150 years, military and civilian energy infrastructures have developed around a small group of liquid fossil fuels, making these desirable candidates for next-generation energy systems owing to a pre-existing logistical infrastructure. These "logistics fuels" include military diesel fuels (e.g. JP-5, JP-8) and civilian liquid fuels (e.g. commercial diesel and gasolines). Alternative fuels include methanol, ethanol, methane, propane, butane and several biomass-derived fluids. In the light of the broad range of available fuels (discussed in detail in Chapter 8), thermal considerations for several processing routes are presented herein.

7.2
Discussion of Component Processes

7.2.1
Overview

The conversion of liquid fuel mixtures to power requires multiple unique chemical and/or physical operations, often carried out by a network of individual processing units, with each unit nominally addressing one required function (e.g. vaporiza-

Microfabricated Power Generation Devices. Edited by Alexander Mitsos and Paul I. Barton
Copyright © 2009 WILEY-VCH Verlag GmbH & Co. KGaA, Weinheim
ISBN: 978-3-527-32081-3

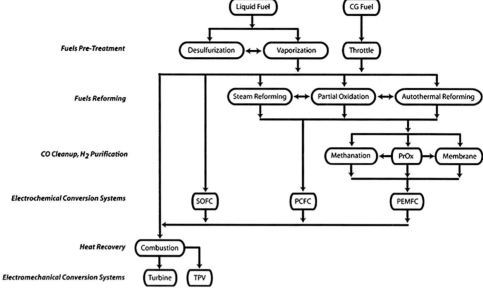

Figure 7.1 Integration routes for conversion of liquid fuels to electrical power in portable devices.

tion, heat recovery, combustion). Figure 7.1 presents a simplified selection tool for fuels and energy conversion system(s) assembly from a suite of functions to comprise an overall portable power system. Table 7.1 summarizes the thermal design constraints (heat duties, operating temperatures) and other function-specific issues associated with each operation. A brief review of each function and its place in the design of heat-integrated portable power systems is provided below. A more detailed discussion of these components is presented in Chapter 3.

7.2.2
Fuels Pre-treatment

Following the selection tool presented in Figure 7.1, fuels pretreatment is considered to be any chemical and/or physical modification of the fuel, as stored, into a form and/or composition suitable for reforming or direct combustion components. The present discussion is focused upon the two most commonly required pre-treatment components, desulfurization and vaporization.

7.2.2.1 Desulfurization
Current petroleum-derived logistics fuels with pre-established infrastructures (e.g. JP-8, diesels, gasoline) typically have total sulfur contents ranging from 300 to 3000 parts per million (ppm). Reduction of sulfur levels is critical to ensure stable operation of catalytic processes (e.g. reforming, electrochemical cells) within

Table 7.1 Summary of components for portable-power devices.

Function	Function Class	Operating Temperature	Heat Duty (kJ mol⁻¹)	Notes
Vaporization	Pre-treatment	150–350 °C	35–85	
Desulfurization (gas-phase)	Pre-treatment	200–400 °C	40–140	
Steam reforming	Fuels reforming	See Table 7.3	See Table 7.3	
Partial oxidation	Fuels reforming	See Table 7.3	See Table 7.3	
Autothermal reforming	Fuels reforming	See Table 7.3	See Table 7.3	
Preferential oxidation	CO clean-up	150–200 °C	−280	
Water-gas shift	CO clean-up	120–220 °C	+41	
Gas separation	CO clean-up	300–800 °C	Negligible	May be coupled with reforming or CO clean-up
Heat exchanger	Heat recovery	Variable	Not Applicable	
Combustion	Heat production	>RT	Variable	Operable from combustibles
PEMFC	Power gen.	70–120 °C	Variable	Requires high-purity H_2 (CO < 10 ppm)
TPV	Power gen.	>800 °C	Variable	Operable from combustibles
Engine	Power gen.	150–600 °C	Variable	Operable from combustibles
IT-SOFC	Power gen.	400–800 °C	Variable	Operable from H_2
SOFC	Power gen.	800–1000 °C	Variable	Operable from combustibles

the portable power system [1]. Gas-phase sulfur removal is accomplished via a hydrodesulfurization reaction followed by removal of hydrogen sulfide by adsorption on metal oxides [1]. Reaction of hydrogen gas and sulfur-containing thiols to yield hydrocarbons and hydrogen sulfide is performed between 200 and 400 °C

Table 7.2 Summary of fuels and vaporization properties.

Liquid Fuel/Oxidant	ΔH° (kJ mol^{-1})	T^{vap} @ 1atm (°C)
Water	40.7	100
Methanol	37.4	65
Ethanol	42.0	78
Butanol	51.0	118
Gasoline (n-octane)	41.6	126
Diesel (n-tetradecane)	71.3	250

[1, 2]. Subsequent adsorption of hydrogen sulfide on zinc oxide is typically carried out at temperatures between 300 and 400 °C [1]. Heats of reactions for gas-phase hydrodesulfurization typically range between 40 and 140 kJ mol^{-1}. Alternatively, liquid-phase extraction of sulfur-containing components via novel copper- and silver-containing zeolites can be performed prior to fuels vaporization [3, 4].

7.2.2.2 Vaporization

Vaporization of liquid fuels and optional oxidants (e.g. water) is required for gas-phase desulfurization, fuels reforming and use in power-conversion components. Table 7.2 summarizes the boiling points and heats of vaporization for several liquid fuels under consideration for use in portable power systems. Multiple authors have reported complete vaporization of heat transfer fluids at degrees of superheating of 50–100 °C [5, 6]. Microscale fuel vaporizers reported to date in portable power systems have been driven by heat supplied via combustion or exothermic reforming at comparable degrees of superheating [7–14].

7.2.3
Power Production Components

Several methods for harnessing power from fuels have been developed at the portable-scale. Power generation methods include direct combustion of fuels followed by conversion of heat energy to electrical power via heat-engines, thermophotovoltaics, thermoelectrics and electrochemical cells. With one exception, all of these power generation systems rely upon gas-phase reactions and handling of gaseous fuels. For the special case of the low-temperature direct fuel cell (e.g. direct-methanol, direct-formic acid, direct-ethanol), pre-treatment and reforming of the fuels is not required, rendering thermal management considerations negligible in comparison to other transport issues – specifically, gas and water management, mass-transfer limited current densities and fuels crossover [15, 16]. Detailed discussion of power production components is presented in Chapters 3, 4 and 5.

7.2.4
Fuels Reforming

Fuel cell components have unique requirements of fuels composition and quality, which may necessitate varying degrees of upstream fuels reforming. Low-temperature proton-exchange membrane fuel cells (PEMFC) require high-purity hydrogen with carbon monoxide contents below 10 ppm [17, 18], while hydrogen-driven intermediate-temperature fuel cells utilizing proton-conducting perovskites are less susceptible to carbon monoxide and can also be operated from unreformed fuels via internal indirect reforming [19, 20]. Solid oxide fuel cells employing oxygen-conducting electrolytes are capable of utilizing carbon monoxide and hydrogen fuels directly [21, 22], although pre-reforming of fuels to a mixture of carbon monoxide, hydrogen and methane is often desirable to mitigate coke formation within the cell [23, 24].

Several reforming routes are available for converting hydrocarbon fuels to hydrogen and/or carbon monoxide and methane, and multiple comprehensive reviews of catalyst [25] and reforming system [26, 27] design have been presented in the literature. Table 7.3 provides a summary of representative heat duties, temperatures and catalysts for both wet (steam) and dry (partial) oxidative reforming of some common hydrocarbon fuels. A detailed discussion of portable fuels reforming is presented in Chapter 3. Hydrogen clean-up, to remove carbon monoxide and methane contaminants, can be subsequently accomplished via water-gas shift [25, 27], preferential oxidation [25, 54], or methanation [55]. Hydrogen purification via palladium membranes [56–58] or mixed-conducting perovskites [59, 60] can also be employed to obtain high-purity hydrogen; the latter ceramic membranes have the additional advantage of greater resistance to corrosion by carbon monoxide and oxidizers. Alternatively, composite permselective-catalytic membranes comprised of catalytic films overlying dense permselective films can be employed for greater corrosion resistance and integrated reforming / purification applications [28].

7.2.5
Thermal Integration, Combustion

Each component has its own unique heat duty and operating temperature which dictates the heat quality of the exothermic components; these heat duties and qualities must be considered in the design of thermally integrated power systems. For example, high quality (800–1000 °C) heat produced by a solid oxide fuel cell (SOFC) can be utilized by downstream turbine systems in a SOFC–turbine hybrid power system [61], or to drive upstream pre-reforming and/or fuel vaporization. Alternatively, high-quality heat can be provided within the portable power system by catalytic combustion of fuel. The use of combustion-heating of fluids in regenerative combustion systems has received substantial treatment in the literature [62, 63]. More recently, the use of combustion heat to drive endothermic fuels

Table 7.3 Summary of heat duties, reported operating temperatures and catalysts for fuels reforming.

Fuel	Steam Reforming ΔH^{25C} (kJ mol^{-1})	Temperature (°C)	Catalysts	Partial Oxidation ΔH^{25C} (kJ mol^{-1})	Temperature (°C)	Catalysts
Methanol [28, 29, 30–33]	49.5	250–300	Al–Cu–Fe Pd/ZnO CuO/ZnO/Al$_2$O$_3$	−192	250–350 200–300 350–525	Cu–Zn/Al$_2$O$_3$ Cu–ZnO/Al$_2$O$_3$ LaNi$_{0.95}$Co$_{0.05}$O$_3$/Al$_2$O$_3$
Ethanol [34–39]	174	250–300 400–500 400–700	Raney Cu–Ni, Rh/CeO$_2$–ZrO$_2$ Rh/Ce$_{0.8}$Zr$_{0.2}$O$_2$	−552	260–380 375–450 500–800	CuO/CeO$_2$ Ni–Rh/CeO$_2$ Rh–Ce/Al$_2$O$_3$
Butane / Propane (LPG) [40–43]	370–490	300–400 350–550	Ni/δ-Al$_2$O$_3$ Pd/CeO$_2$	−1000—1449	350–475 400–500	Pt–Ni/δ-Al$_2$O$_3$ Pt/δ-Al$_2$O$_3$
Gasoline [44–48]	846	580 580 700	Ni/CeZSM-5 Ni–Re/Al$_2$O$_3$ NiSr/ZrO$_2$	−3000	500–750 675–775 1100	Ni–Pd/g-Al$_2$O$_3$ Rh–MgO/CeO$_2$–ZrO$_2$ Rh/α-Al$_2$O$_3$
Diesel [49–53]	~1600–2300	600–900 885–950	Pd/ZrO$_2$ Mo$_2$C	~5000—5600	350–650 600–700 500–800	Pt/CeO$_2$ Rh–Ce/Al$_2$O$_3$ La$_y$Ni$_{1-y}$Ce$_x$Fe$_{1-x}$O$_3$

reforming in heat exchanger reactors has emerged as a viable means of process intensification [64, 65].

7.3
Integration Schemes, Methods and Dimensions

From the above discussion, it can be seen that the design of the overall portable power device requires integration of several unique functions, each with its own unique thermal criteria, within a single volume. Two contrasting strategies are identified for accomplishing this goal: (i) construction of several individual components, each fulfilling a single chemical and/or physical function, and subsequent integration via fluidic connections (process integration); or (ii) construction of a single-component device containing all chemical and/or physical functions within one monolithic structure (component integration). The former strategy is well known to the chemical engineer, as it remains the basis for conventional process design, while the latter strategy has gained in popularity via two decades of process intensification research. While the former strategy allows *energy integration* strategies (i.e. directing of heat from exothermic to endothermic processes via convection) already developed for conventional chemical process design and optimization, the latter requires far more complex manipulation of heat conduction within the monolithic structure in one or more dimensions via *process intensification*.

7.3.1
Horizontal (Zero-Dimensional) Integration

Horizontal integration denotes the parallel fabrication of multiple individual components, followed by interconnection via fluidic and/or electrical "breadboarding" to comprise the integrated system. In this integration strategy, each component is first designed and fabricated as a stand-alone device and is often assumed to be thermally isolated from the rest of the system. The overall system can thus be fabricated and designed in a manner analogous to conventional plant design, where thermal management and energy integration are accomplished by connection of utility (steam, water) lines. The most immediate advantages of this strategy are:

- Applicability of established plant design and energy integration [66] methods to the design of microscale systems.
- Facile maintenance of system, via removal/replacement of individual failed components, as opposed to replacement of entire system.
- Freedom to design each component independently of its fellow components for maximum effectiveness and minimum design costs.
- Direct control over operating temperatures and heat flow between components via routing of heat-transfer fluids.

There exist multiple examples of microscale designs for effecting horizontal integration in chemical analysis and organic synthesis. Ehrfeld and coworkers

[67, 68] have developed a modular "building block" packaging system capable of housing a vast array of compact, efficient single-functional microsystems (e.g. heat exchangers, mixers, reactors). Fluidic connections provided by the packaging system enable straightforward use of plant design concepts and process intensification strategies utilizing fluid convection to direct heat between individual processes. Because each component is tailored for a single, robust function, a wide variety of microchemistry processes can be constructed from a relatively small assortment of components (Figure 7.2a). Lobbecke and coworkers [69, 70] have likewise developed their own system for integrating microchemical subsystems within a fluidic breadboard, which provides utility connections (heating fluids, coolant fluids) in addition to feed/effluent connections to each component device. These subsystems are, in many cases, comprised of multiple elements (e.g. mixer, heat exchanger and reactor in a single component) to achieve a better balance between modular design flexibility and the benefits of process intensification and reduced modular system complexity (i.e. number of elements). The authors employ a hexagonal sealing face for connecting elements to the breadboard; this hexagonal face design allows up to six fluidic connections to each element to realize complex integration schemes. Additionally, this fluidic breadboard also provides simple electrical connections between the breadboard and individual chips, for the purpose of temperature monitoring, and optical ports for visual and/or spectroscopic monitoring of reactions within individual units (Figure 7.2b). Palo and coworkers have reported on a breadboard system for portable power applications, comprised of several previously developed single- and multi-functional components [71, 72]. In this system, individual components are combined with balance-of-plant components to comprise a compact portable power system, with emphasis upon energy intensification via conventional plant design methods (Figure 7.2c).

The primary challenges to thermal management in horizontal integration of components are: (i) process design optimization for energy integration, analogous to challenges in conventional plant design [66, 73]; and (ii) addressing heat conduction from individual components to the breadboard platform via packaging. This second item is unique to microsystems, and represents a significant challenge in the final design and assembly of any portable power system. However, before discussing packaging methods, one must consider strategies for process intensification via vertical integration strategies to improve component efficiencies and reduce overall system size and complexity.

7.3.2
Component Integration Strategies

Component integration strategies combine multiple functions within a single integrated component. In this integration stategy, heat transfer from exothermic processes to endothermic processes occurs not only via convection of feed, effluent and heat-transfer fluids, but by direct conduction of heat through the solid medium. Appropriate selection of one-, two- and three-dimensional layouts and/or materials patterning is required to maintain the target operating temperatures of each

(a)

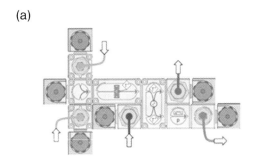

(b)

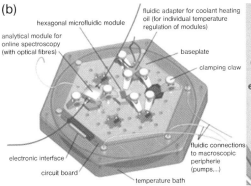
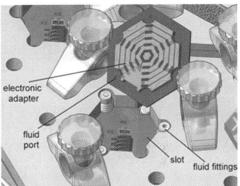

(c)

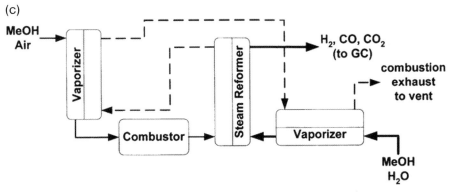

Figure 7.2 Process integration schemes for microchemical systems: (a) Ehrfeld Mikrotechnik BTS GmbH Modular Microreaction Technology System. Each component presents a single, highly efficient function; components are connected to comprise an overall system following conventional process synthesis strategies (reprinted with permission, from MRT-Catalog 2007); (b) Illustration of the modular microreaction system, from [69, 70] (reprinted with permission from John Wiley and Sons); (c) Schematic of bread-board fuel processor reported by Palo and coworkers [71]. Energy integration achieved by combustion-based pre-heating of steam/ fuel supply to reformer and recovering waste reformer heat contained in effluent gas through fuel vaporizers (reprinted with permission from Elsevier).

function, maximize thermal efficiency and minimize heat losses via conduction to packaging and/or the surroundings. In the following section, strategies for directing heat conduction via one-, two- and three-dimensional architectures for realizing integrated components are discussed.

7.3.2.1 Cartesian (One-Dimensional) Integration

The most common means for integrating multiple functions in a single component involves layering or stacking of 2D arrays of identical microchannels within a plane orthogonal to the direction of flow, such that a 1D progression of functions is accomplished within the microdevice (Figures 7.3 and 7.4) by stacking flat plates. Appropriate selection of the 1D stacking layout or order enables the creation of significant thermal gradients perpendicular to the direction of flow to facilitate heat integration, provided that materials with sufficiently low thermal conductivities are incorporated within the microreactor stack. Alternatively, use of high thermal conductivity materials allows rapid heat exchange between each functional layer, resulting in thermal equilibration between components and isothermal operation.

Several reports of isothermal multifunctional components employing 1D Cartesian integration designs have appeared in the literature over the past decade. Park and coworkers reported on the performance of a microreactor component comprising fuel vaporization, combustion and steam reforming, integrated vertically via stacked metal microchannel plates [8] (Figure 7.3a). The resulting system was demonstrated over a range of operating temperatures of 220–280 °C, with

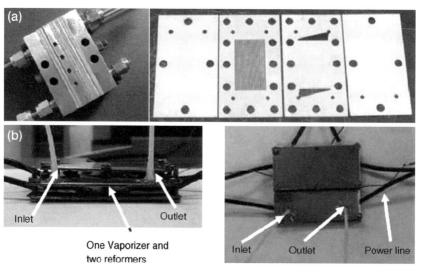

Figure 7.3 Examples of 1D integration of multiple functional layers resulting in isothermal operation: (a) machined metal plate stack comprising vaporizer, heat exchanger, steam reformer and catalytic combustion, from [8], and (b) machined silicon plates stacked to comprise vaporization and steam reforming (with on-chip heating via externally applied voltage, from [9]. Images reprinted with permission from Elsevier.

temperature variations across the entire integrated stack of less than 6 °C. Kwon and coworkers [9] reported an integrated component comprising fuel vaporization and steam reforming fabricated in silicon and operated at isothermal temperatures of 250–300 °C. In both cases, the systems were constructed from high thermal conductivity materials (silicon, metal) with sub-millimeter heat conduction pathways, which resulted in thermal equilibration between functional layers and isothermal operation.

As discussed in Section 7.2, each function has an optimal operating temperature range, such that non-uniform temperatures across the integrated component are desirable. Unique operating temperatures within each functional layer can be maintained by incorporation of insulating barriers between functional layers. Delsmann and coworkers [74] reported on an integrated preferential oxidation–heat exchanger comprised of a vertical stack of microstructured metal plates integrating heat exchangers with preferential oxidation (Figure 7.4a). Unique operating temperatures were maintained for each zone by placing insulating layers between the heat exchanger zones (175 and 125 °C) and the preferential oxidation (165 °C). Shah and Besser [26] presented an integrated reformer consisting of a catalytic packed-bed steam reformer sandwiched between smaller preferential oxidation and combustion microreactors (Figure 7.4b). By incorporating vacuum-sealed gaps between the central steam reforming and the two exothermic reaction zones, operating temperatures of 564, 230 and 150 °C were achieved for the combustion, steam reforming and preferential oxidation zones, respectively. Terazaki and coworkers [11] utilized a similar strategy to realize an integrated reformer combining steam reforming, combustion, preferential oxidation and vaporization in micromachined glass plates. In this design (Figure 7.4c), three separate insulation gaps were provided to separate the coupled combustion–steam reforming region from the two unique vaporization regions; subsequent vacuum packaging of the completed device allowed operation of these three regions at temperatures of 280, 180, 140 and 110 °C, respectively, during preliminary tests utilizing on-chip electrical heating in lieu of catalytic operation [11].

7.3.2.2 Radial (One-Dimensional) Integration

A primary limitation of Cartesian stacking of functional layers is that heat loss to the surroundings is identical at either extreme of the vertical stack, that is, substantial insulation must be provided to the surface of the highest temperature region. These losses are compounded by further losses through the external boundaries of each plate or layer. In the light of these challenges, radial integration of functional layers has been proposed as a means of mitigating these losses while removing a substantial insulation requirement of the packaging [75].

By utilizing a tubular architecture with radial layering of functional zones, heat losses from internal layers are restricted to losses to packaging at the fluid inlet and outlet. The requirement of symmetry at the centerline of the radial "stack" supports a self-insulating design, allowing placement of the lowest temperature functions in the outermost layer, in direct contact with external packaging and insulation to minimize heat losses from the system. As is the case in Cartesian integration strategies, maintenance of unique operating temperatures within each

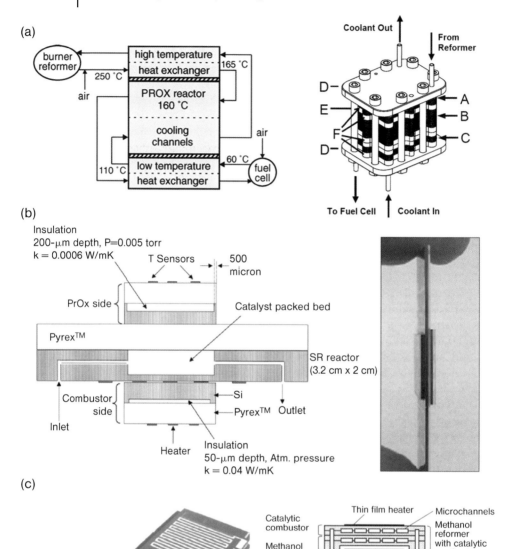

Figure 7.4 Examples of 1D integration of multiple functional layers with unique operating temperatures in integrated stacked-plate components. (a) stainless-steel plates coupling heat exchangers with preferential oxidation; component couples with separate burner/reformer and fuel cell components via horizontal integration to comprise overall system, from [74]; (b) six-layer silicon microreactor with central steam reformer integrated with preferential oxidation and combustion on front- and back-side, from [26]; (c) micromachined glass plates coupling methanol reforming, combustion, preferential oxidation with vaporization, from [11]. Images reprinted with permission from Elsevier.

layer can be achieved by incorporating insulation layers between functional layers. It remains a significant fabrication challenge to develop efficient means for constructing multiple tubes-in-shells systems at the microscale or, equivalently, distributed bundles of microchannels, while maintaining series and parallel connectivity between individual channels.

7.3.2.3 Complex (Two- and Three-Dimensional) Integration

In lieu of simple Cartesian or radial stacking of functional layers, more complex 2D and 3D layouts can be realized via existing micromachining methods. These architectures allow the creation of complex 2D and 3D thermal gradients within a single component, while incorporation of multiple construction materials allows further manipulation (either by insulation or heat shunting) of thermal gradients within the component.

An early example of this strategy was presented by Arana and coworkers [76], shown in Figure 7.5e. This device, fabricated from silicon using Si-MEMS technologies, integrates an isothermal reaction volume (separated into two reaction compartments) with a four-fluid heat exchanger and an external packaging frame. The isothermal reaction volume is suspended upon thermally insulating silicon nitride tubes, resulting in a thermally isolated "hot-box" wherein heat produced by hydrogen combustion in one compartment drives endothermic reforming of ammonia in the other. Vacuum-packaging of the assembled device ensures minimal convective heat losses from the isothermal "hot-box" region, while silicon nitride feed/effluent tubes minimize heat conduction from the "hot-box" region to the packaging region. Heat exchange between hot gas effluent and cold gas feed in the silicon nitride tubes is enabled by placement of transverse silicon conduction bars at regular intervals along the heat exchanger length. This design thus represents a combination of process intensification and energy integration, both accomplished by heat conduction through the microdevice (Figure 7.5).

A second example of complex thermal management is the use of free-standing membranes to create 2D thermally isolated regions within the component. Free-standing silicon nitride, silicon oxide and composite membranes [56, 57, 77, 78] provide a thermally isolated platform for multiple high-temperature applications (Figure 7.6). The low thermal mass associated with the thin (~0.2–20 µm) film allows rapid heating and cooling, while the low thermal conductivity minimizes heat loss via conduction to the silicon and/or glass substrate. However, this architecture limits the size of the functional region to the high-temperature membrane surface. The size and durability of the functional membrane is in turn limited by its mechanical stability.

Manipulation of thermal gradients in single- and multi-functional components can also be accomplished by incorporation of additional patterned layers of high thermal conductivity materials in order to redirect heat flow across the component. The use of copper-metal "thermal spreaders" in silicon microreactors was demonstrated to be capable of hot spot removal for combustion reactions [80] and for routing combustion heat to an integrated thermoelectric layer for portable power generation [81].

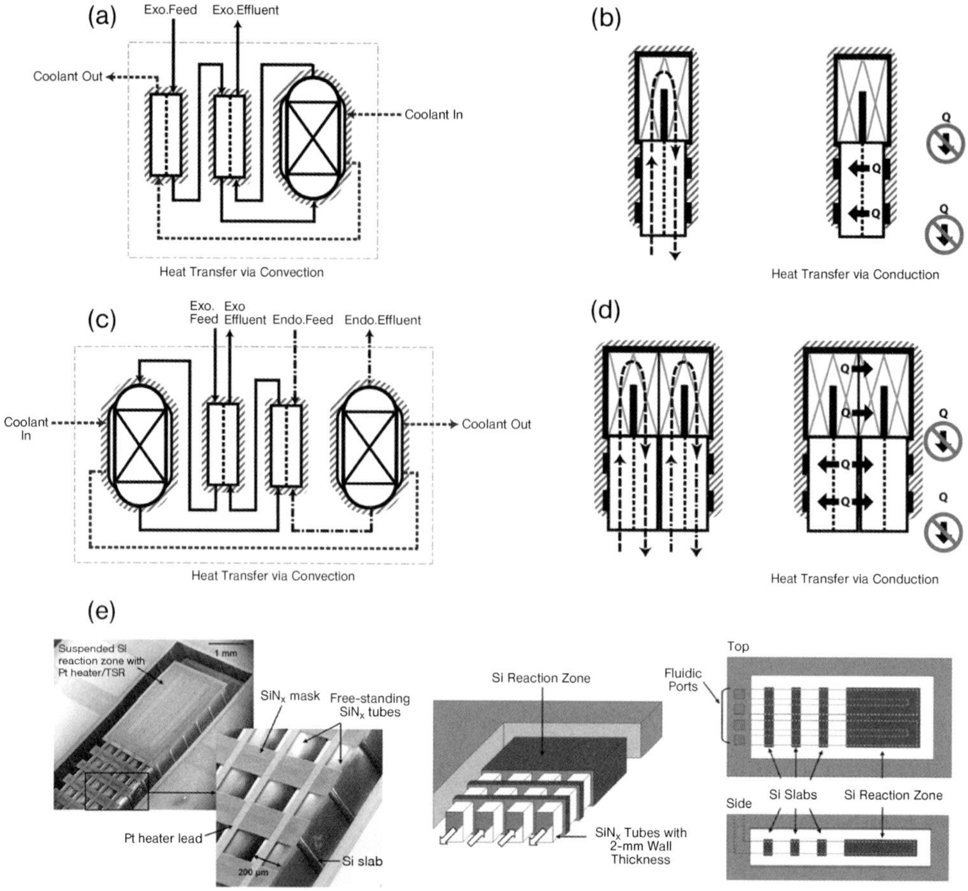

Figure 7.5 Energy integration in conventional chemical process design via horizontal integration of several components vs. process intensification in vertically integrated microdevices, from [76]: (a) energy integration for a single exothermic chemical reactor with feed–effluent heat exchanger. Additional heat recovery accomplished via coolant-feed heat exchanger. (b) Equivalent system achieved in "hot-box" architecture [76]. (c) Energy integration for a single exothermic chemical reactor coupled with a single endothermic chemical reactor. Additional energy intensification accomplished by routing heat transfer fluid between the two reactors. (d) Process intensification in "hot-box" architecture [76]. (e) Thermally isolated hot-box microdevice, from [76]. Images reprinted with permission from John Wiley and Sons.

7.4
Materials of Construction

In conjunction with the selection of component geometry, the selection of construction and insulation materials supports thermal management in integrated components. Materials selection is critical for manipulation of thermal profiles

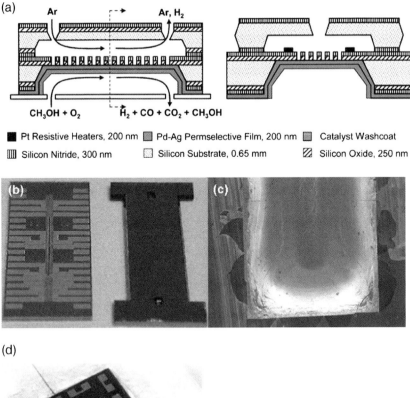

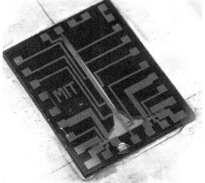

Figure 7.6 Thermally isolated micromembrane designs: (a) silicon oxide/nitride composite membrane for thermally-isolated hydrogen extraction from fuels, cross-section schematic, (b) view of component layers, (c) catalyst coating, from [8]; (d) silicon nitride membrane reactor for catalytic reforming, from [79].

and heat conduction pathways within integrated components, as well as the overall portable power device. Simultaneously, materials selection must be tempered with other practical concerns, including availability and applicability of machining methods, ease of manipulating and/or introducing catalytic or surface functionalities, and mechanical limitations. In this section a discussion of the primary

Table 7.4 Summary of the properties of common construction materials.

Material	Density (gm cm^{-3})	Thermal Conductivity (W m^{-1} K^{-1})	Coefficient of Thermal Expansion (°C^{-1})	Melting Point (°C)
Silicon	2.3	150	8.2×10^{-4}	1380
SiO$_2$	2.2	1.4	4.5×10^{-7}	1680
Pyrex 7740	2.2	1.1	3.3×10^{-6}	1250
Si$_x$N$_y$ (CVD, PECVD)^a	3.0–3.3	1–5	3.3×10^{-6}	1900
Air @ 1 bar	1.3×10^{-3}	$\sim 2 \times 10^{-2}$	N.A.	N.A.
Air @ 2×10^{-3} mbar (1×10^{-3} torr)	0.8×10^{-7}	$\sim 3 \times 10^{-4}$	N.A.	N.A.
Copper	8.9	380	1.7×10^{-5}	1080
S.S. 316L	8.0	20	2.0×10^{-5}	1370
SiC	>3.10	125	5.2×10^{-6}	2730
α-Al$_2$O$_3$	3.8	25	8.4×10^{-6}	2000

a   Thermal properties vary depending upon nitride stoichiometry, deposition methods, porosities.

materials for construction employed for fabrication of high-temperature portable power components is provided. The thermal properties for materials discussed in this section are summarized in Table 7.4.

7.4.1
Silicon-Based Materials

The use of single-crystal silicon as a substrate for the machining of complex, 2D structures via established integrated circuits (IC) and microelectromechanical systems (MEMS) fabrication methods was first reported by researchers at MIT and the DuPont de Nemours Company [79, 82]. Since then, several single- and multi-functional components for portable power generation have been demonstrated in silicon-based architectures for solid oxide fuel cells [29, 83, 84], proton-exchange membrane fuel cells [85–87], single-stage fuels reforming [24, 54, 76, 79, 88], and multistage fuels reforming [26, 28, 76]. Likewise, silicon oxide (quartz) or sodium-glass (e.g. pyrex) wafers can be used as substrates for IC-MEMS fabrication of single- and multi-functional components [11]. In both cases, stacking of individual plates or slabs into integrated components can be accomplished by anodic-bonding or fusion-bonding of sodium-glass, quartz and/or silicon layers [89, 90].

Single crystal silicon is a strong, low-density material with a thermal conductivity comparable to that of brass, allowing production of low-weight, rapidly cooled or heated components capable of operating under isothermal or near isothermal conditions. The high thermal conductivity of silicon has made silicon-based micro-devices especially suited for safely handling highly exothermic chemistries without temperature run-away [91–94], and for evaluating catalytic activity and reaction kinetics under isothermal conditions [95, 96]. For applications in which high-temperature operation, with or without internal thermal gradients, is desired, with minimal heat losses, the thermal conductivity of silicon can be augmented via vacuum packaging and/or integration of multiple thin-film materials via IC-MEMS technology.

The incorporation of vacuum-sealed thermal barriers within silicon devices to block heat conduction from one functional region to another has been demonstrated by several authors. Besser and coworkers have demonstrated a means for introducing vacuum gaps between single silicon layers in a multifunctional reforming stack [26], with the additional flexibility of being able to tune the vacuum level within each insulating gap to optimize thermal conductivities and resulting thermal profiles [75] (Figure 7.4b). A similar strategy, wherein open gaps were included between individual micromachined glass layers to induce thermal gradients, was demonstrated by Terazaki and coworkers [11]; vacuum-packaging of the resulting multifunctional component further reduced the thermal conductivity of these insulation barriers (Figure 7.4c).

Silicon-based processes inherit immense flexibility in the introduction of 2D patterned thin-films, either by physical or chemical vapor deposition methods. Silicon nitride, introduced via chemical vapor deposition in thicknesses ranging from nanometers to tens of microns, provides a flexible means of introducing thermal and/or electrical barriers. Silicon nitride can be employed as a thermal barrier film, or for the construction of structural components capable of providing thermal isolation between a high-temperature region and the silicon substrate. As discussed in Section 7.3, silicon nitride has been used to build free-standing thermally isolated membranes (Figure 7.6) and thermally insulating supporting struts to isolate high-temperature reaction zones (Figure 7.5). Silicon oxide, deposited on the silicon substrate by chemical or physical deposition methods or grown on the silicon substrate via oxidation, provides a second material for introducing both electrical and thermal insulation. Silicon oxide films have also been employed for creation of thermally isolated membranes [56] and thermally isolating support structures for isolation of high-temperature reaction zones [97].

Thin films of aluminum, copper or gold (commonly patterned on silicon in the IC-industry) allow manipulation of thermal gradients by presenting 2D patterned layers capable of redirecting heat flow. To date, this capability has been utilized for the design of "thermal spreaders" which enable the management of excess heat generated by exothermic reactions [80, 81], as discussed in Section 7.3.

7.4.2
Metals

Metal construction materials provide design flexibility in chemical compatibility at low cost relative to silicon micromachining and reduced materials processing relative to ceramics. Metals employed for portable power components have thermal conductivities approaching that of silicon, which can be mitigated by incorporation of insulating plates or air gaps [74].

Metal sheets may be machined with complex 2D patterns by several methods, including conventional milling [98, 99], electric-discharge machining [100, 101] or lithography-based chemical etching [8, 10, 74, 102]. Stacked systems of plates can then be sealed by brazing [99] or by compression between two packaging layers containing external fluidic connections [8, 74,98, 99, 102]. Materials of construction are typically stainless steels [10, 74, 98, 99, 102], but other metals have also been employed [100, 101].

The thermal conductivities of metals (Steel, Ni, Al, Cu) typically employed for microreactor manufacture are comparable to that of silicon and present the same thermal considerations. To maintain thermal gradients across the microdevice, thermal insulation layers may be incorporated in the metal-plate stack. Delsman and coworkers [74] have employed air-filled cavities and low thermal conductivity insulation materials between individual plates to present thermal barriers within integrated microdevices.

7.4.3
Ceramics

Ceramic microreactors can be fabricated by several methodologies. Knitter and coworkers [103] have detailed a robust methodology in which 3D molds are first fabricated by stereolithography, followed by injection molding and sintering of the ceramics. This methodology has been demonstrated for the production of several complex geometries in α-Al$_2$O$_3$ and is expected to be extendable to several other ceramic materials [103]. Alternatively, pre-fired or sintered ceramic sheets may be machined, stacked and sealed in a fashion similar to silicon- and metal-stacked microdevices. Schmitt and coworkers demonstrated a procedure in which α-Al$_2$O$_3$ sheets are tape-cast onto organic support tapes and laser cut to the desired patterns; the complete ceramic microdevice is then fabricated by stacking, sealing and sintering individual plates [104]. Meschke and coworkers detail methods for

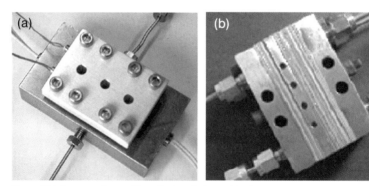

Figure 7.7 Compression-sealing of microdevices: (a) silicon-fabricated microreactor sealed in chuck assembly, from [95]; (b) compression-sealed stainless-steel microreactor stack with packaging layers containing tube fittings and heater ports, from [8]. Images reprinted with permission from Elsevier.

sintering stacks of pre-fired SiC plates which have been machined by either conventional milling or laser machining [105].

Ceramic materials offer potential advantages over silicon-based systems in terms of thermal management, catalyst introduction and materials functionality, many of which remain to be fully exploited. Ceramics are available over a wide range of thermal conductivities, down to values two orders of magnitude lower than silicon or metals. The use of low thermal conductivity ceramics substantially reduces axial conduction losses in heat exchanger and process intensification systems [106, 107]. The porosity and density of ceramic materials may be manipulated both by materials selection and by sintering or firing procedures, potentially offering greater ease and flexibility for integrating mass-transfer applications. Tunable porosities combine with rough surfaces to facilitate catalytic washcoat adhesion and catalyst impregnation. Lastly, the potential to construct components and/or functional layers from electroceramics for gas purification or fuel cell applications remains to be exploited in ceramic portable power systems.

7.5
Packaging Strategies

Upon completion of individual single- and/or multi-functional components, assembly of components into the overall portable power system requires packaging, or the provision of fluidic and electrical connections between individual components, either directly or via fluidic breadboards. This section discusses the different packaging methods employed in microchemical systems to provide fluidic connections, while addressing thermal considerations (operating temperatures, heat losses) for each.

7.5.1
Compression Sealing

The most straightforward means of ensuring both fluidic and electrical connections between the packaging assembly and stand-alone microcomponents is by compression sealing of components directly to a packaging layer, typically either a fluidic breadboard [70] or chuck assembly [93, 95, 108] (Figures 7.2 and 7.7). This in turn allows conventional tubing fittings to be incorporated into the packaging layer, alleviating the need for direct fluidic connections to micromachined components. Alternately, Gonzalez and coworkers [113] have demonstrated self-aligning "finger microjoints" capable of clamping two separate microdevices together with chip-to-chip connections via a compression gasket, without the need for external chuck assemblies; in this system, direct connection of plumbing to microfabricated components is still necessary. In all cases, the use of compression seals allows reversible packaging of components for greater design flexibility and component replacement, while enabling use of chemical-resistant seals (e.g. fluorinated polymer gaskets, including Viton).

Compression sealing presents two significant thermal design constraints on the overall system design. First, compression gaskets must be capable of maintaining seals from ambient temperature up to the microdevice operating temperature, which limits the use of fluorinated polymer gaskets to applications below 200 °C and presents substantial challenges for high-temperature reforming. Secondly, large interfacial surface areas between microdevice and packaging, necessary for compression, facilitate substantial heat losses from high-temperature microdevices. In applications where temperature control of the components is critical, this facilitates indirect heat removal and/or addition to the microdevice via cooling or heating of the packaging layer [8]. In portable power systems, substantial heat losses from individual components to the packaging layer makes insulation of the packaging layer critical to minimizing heat losses, and may lead to thermal equilibration between separate components through the packaging layer.

7.5.2
Direct Fluidic Connections

The alternative to providing fluidic and electrical connections to a packaging layer via compression-sealing devices is to directly attach tubing and wiring to the microdevice. Several methods are available, depending on the target operating temperatures and the materials of construction. In all cases, heat losses from components are substantially reduced, as the interfacial contact area between the packaging (tubing) and the microdevices is minimized. Examples of several direct fluidic connections reported to date are shown in Figure 7.8. Heat conduction losses through tubing can be further mitigated by employing low thermal conductivity materials such as quartz or glass [114].

7.5.2.1 Micromachined Mechanical Couplers
Fluidic seals between individual microdevices, fluidic breadboards and/or tubing can be accomplished using micromachined couplers integrated into component devices. Meng and coworkers have developed machined silicon compression couplers capable of both chip-to-chip and tubing-to-chip seals [115]. Zero-dead volume receptacles machined in silicon microdevices, capable of receiving capillary tubing, have been demonstrated by Gray and coworkers [116]. A similar strategy in which heat-shrink tubing is used to seal rigid tubing to on-chip receptacles has been demonstrated by Pan and coworkers [109]. These fabrication strategies rely upon matching expansions of both tubing and substrate to prevent loss of sealing, which limits practical operating temperatures.

7.5.2.2 Adhesive Seals and Brazed Seals
Low-temperature adhesives can be employed to seal tubing to fluidic ports on microdevices [92]. This methodology is popular in low-temperature and low-pressure applications, given the simplicity and low cost of packaging. In portable power applications, low-temperature epoxies can be replaced with suitable ceramic adhesives [90]. Direct brazing or welding of tubing to devices provides a robust, high-temperature packaging methodology. The use of glass brazes for sealing metal

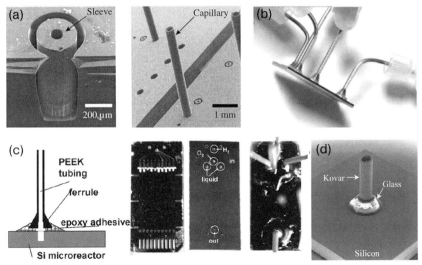

Figure 7.8 Direct fluidic packaging of microdevices:
(a) mechanical couplers, from [109] (reprinted with
permission from IEEE); (b) solder-joint packaging, from [110],
(reprinted with permission from RSC Publishing); (c) epoxy-
joint packaging, from [111], (reprinted with permission
from Elsevier); (d) glass-brazing, from [112] (reprinted with
permission from IEEE).

tubing to silicon microengines has been demonstrated by Peles and coworkers
[112].

7.5.2.3 Solder-Joint Sealing

Fluidic connections between microdevices can be also be accomplished by employing solder joints. Solder joints have been demonstrated at pressures up to 200 bar
at temperatures from −78 to 160 °C and the resulting all-metal fluidic connections
are compatible with a large number of solvents [117]. This is accomplished by
patterning metal (typically copper) bonding pads around fluidic ports on the
microdevice and employing pre-formed solder seals for direct attachment of metal
tubing to individual devices [110, 111], or for connecting individual devices to an
integrated platform [117]. While this bonding methodology is quite promising for
low-temperature applications such as organic synthesis, operating temperatures
are limited to below 200 °C to prevent re-flow of solder and loss of fluidic connections [117].

7.6
Summary

Thermal management of overall portable power processes, by appropriate selection of materials, geometries and packaging, both for each system component and

for integrating the resulting network, is critical to the realization of efficient portable power systems. To fully realize efficient portable power systems, all three aspects of thermal management (component integration, packaging and materials selection) must be considered in tandem. This chapter has attempted to provide an overview of the myriad challenges for each aspect of thermal management, and to provide a review of current solutions and strategies for resolving these challenges.

Acknowledgements

Support for this effort was provided in part by a Junior Faculty Summer Fellowship awarded by the University of Connecticut, a DuPont Young Faculty Grant awarded by the DuPont de Nemours Company, and an Office of Naval Research Young Investigator Grant (ONR # N000140710828). The author gratefully acknowledges a three-year appointment at the Massachusetts Institute of Technology (2002–2005) under the direction of Professor Klavs F. Jensen, which provided the many collaborations and experiences that made this chapter possible, funded by the DoD Multidisciplinary University Research Initiative program administered by the U.S. Army Research Office under Grant DAAD19-01-1-0566U.S. The author thanks the editors (Paul Barton and Alex Mitsos) for their useful comments and suggestions in the development of this work.

References

1 Lampert, J. (2004) Selective catalytic oxidation: a new catalytic approach to the desulfurization of natural gas and liquid petroleum gas for fuel cell reformer applications. *Journal of Power Sources*, **131**, 27–34.

2 Twigg, M.V. (1989) *Catalyst Handbook*, 2nd edn, Wolfe Press, London.

3 Hernandez-Maldonado, A.J. and Yang, R.T. (2003) Desulfurization of liquid fuels by adsorption via p complexation with Cu(I)-Y and Ag-Y zeolites. *Industrial and Engineering Chemistry Research*, **42**, 123–9.

4 Song, C. and Ma, X. (2003) New design approaches to ultra-clean diesel fuels by deep desulfurization and deep dearomatization. *Applied Catalysis B: Environmental*, **41**, 207–38.

5 Jiang, L. Wong, M. and Zohar, T. (2001) Forced convection boiling in

microchannel heat sink. *Journal of Microelectromechanical Systems*, **10**, 80–7.

6 Kandlikar, S.G. (2002) Two-phase flow patterns, pressure drop, and heat transfer during boiling in minichannels flow passages of compact evaporators. *Heat Transfer Engineering*, **23**, 5–23.

7 Tonkovich, A.L.Y., Jimenez, D.M., Zilka, J.L., LaMont, M.J., Wang, Y. and Wegeng, R.S. (1998) Microchannel chemical reactors for fuel processing, in Proceedings of the 2nd International Microreaction Engineering Symposium (IMRET-2), New Orleans.

8 Park, Y., Yoon, Kim, S. and Eguchi, H. (2005) Production with integrated microchannel fuel processor using methanol for portable fuel cell systems. *Catalysis Today*, **110**, 108–13.

9 Hwang, K. and Kim, A. (2006) Silicon-based miniaturized-reformer for portable

fuel cell applications. *Journal of Power Sources*, **156**, 253–9.

10 Seo, Y., Yoon, P. and Kim, P. (2004) Development of a Micro Fuel Processor for PEMFCs. *Electrochemica Acta*, **50**, 719–23.

11 Terazaki, N., Takeyama, N. and Yamamoto, T. (2005) Development of multi-layered microreactor with methanol reformer for small PEMFC. *Journal of Power Sources*, **145**, 691–6.

12 Kundu, A., Jand, J.H., Lee, H.R., Kim, J.H., Gil, J.H., Jung, C.R. and Oh, Y.S. (2006) MEMS-based micro-fuel processor for application in a cell phone. *Journal of Power Sources*, **162**, 572–8.

13 Pattekar, A.V. and Kothare, M.V. (2004) A microreactor for hydrogen production in micro fuel cell applications. *Journal of Microelectromechanical Systems*, **13**, 7–19.

14 Shah, K., Ouyang, X. and Besser, R.S. (2005) Microreaction for microfuel processing: challenges and prospects. *Chemical Engineering and Technology*, **28**, 303–13.

15 Schultz, T., Zhou, S. and Sundmacher, K. (2001) Current status of and recent developments in the direct methanol fuel cell. *Chemical Engineering & Technology*, **24**, 1223–33.

16 O'Hayre, R.P., Cha, S.-W., Colella, W. and Prinz, F.B. (2006) *Fuel Cell Fundamentals*, John Wiley and Sons, Inc., New York.

17 Song, C. (2002) Fuel processing for low-temperature and high-temperature. Fuel cells. Challenges and opportunities for sustainable development in the 21st century. *Catalysis Today*, **77**, 17–49.

18 Qi, Z., He, C. and Kaufman, A. (2002) Effect of CO in the anode fuel on the performance of PEM fuel cell cathode. *Journal of Power Sources*, **111**, 239.

19 Zhu, B., Albinsson, I., Mellander, B.-E. and Meng, G. (1999) Intermediate-temperature proton-conducting fuel cells – present experience and future opportunities. *Solid State Ionics, Diffusion and Reactions*, **125**, 439–46.

20 Iwahara, H. (1988) High-temperature proton conducting oxides and their applications to solid electrolyte fuel cells and steam electrolyzer for hydrogen production. *Solid State Ionics*, **28-30**, 573–8.

21 McIntosh, S. and Gorte, R.J. (2004) Direct hydrocarbon solid oxide fuel cells. *Chemical Reviews*, **104**, 4845–65.

22 Gupta, G.K., Marda, J.R., Dean, A.M., Colclasure, A.M., Zhu, H. and Kee, R.J. (2006) Performance predictions of a tubular SOFC operating on a partially reformed JP-8 surrogate. *Journal of Power Sources*, **162**, 553–62.

23 Eguchi, K., Kojo, H., Takeguchi,T., Kikuchi, R. and Sasaki, K. (2002) Fuel flexibility in power generation by solid oxide fuel cells. *Solid State Ionics, Diffusion and Reactions*, **152–153**, 411–16.

24 Chen, F., Zha, S., Dong, J. and Liu, M. (2004) Pre-reforming of propane for low-temperature SOFCs. *Solid State Ionics*, **166**, 269–73.

25 Ghenciu, A.F. (2002) Review of fuel processing catalysts for hydrogen production in PEM fuel cell systems. *Current Opinion in Solid State and Materials Science*, **6**, 389–99.

26 Shah, K. and Besser, R.S. (2007) Key issues in the microchemical systems-based methanol fuel processor: energy density, thermal integration and heat loss mechanisms. *Journal of Power Sources*, **166**, 177–93.

27 Holladay, J.D., Wang, Y. and Jones, E. (2004) Review of developments in portable hydrogen production using microreactor technology. *Chemical Reviews*, **104**, 4767–89.

28 Wilhite, B.A., Weiss, S.E., Ying, J.Y., Schmidt, M.A. and Jensen, K.F. (2006) Demonstration of 23wt% Ag-Pd micromembrane employing 8 : 1 $LaNi_{0.95}Co_{0.05}O_3/Al_2O_3$ catalyst for high-purity hydrogen generation. *Advanced Materials*, **18**, 1701–4.

29 Huang, H., Nakamura, M., Su, P., Fasching, R., Saito, Y. and Prinz, F.B. (2007) High-performance ultrathin solid oxide fuel cells for low-temperature operation. *Journal of the Electrochemical Society*, **154**, B20–24.

30 Tsai, A.P. and Yoshimura, M. (2001) Highly active quasicrystalline Al-Cu-Fe catalyst for steam reforming of methanol. *Applied Catalysis A: General*, **214**, 237–41.

31 Chin, Y.-H., Dagle, R., Hu, J, Dohnalkova, A.C. and Wang, Y. (2002) Steam reforming of methanol over highly active Pd/ZnO catalyst. *Catalysis Today*, **77**, 79–88.

32 Purnama, H., Ressler, T., Jentoft, R.E., Soerijanto, H., Schlogl, R. and Schomacker, R. (2004) CO formation/selectivity for steam reforming of methanol with a commercial CuO/ZnO/Al₂O₃ catalyst. *Applied Catalysis A: General*, **259**, 83–94.

33 Alejo, L., Lago, R., Pena, M.A. and Fierro, J.L.G. (1997) Partial oxidation of methanol to produce hydrogen over Cu-Zn-based catalysts. *Applied Catalysis A: General*, **162**, 281–97.

34 Nishiguchi, T., Matsumoto, T., Kanai, H., Utani, K., Matsumura, Y., Shen, W.-J. and Imamura, S. (2005) Catalytic steam of ethanol to produce hydrogen and acetone. *Applied Catalysis A: General*, **279**, 273–7.

35 Kugai, J., Velu, S. and Song, C. (2005) Low-temperature reforming of ethanol over CeO2-supported Ni-Rh bimetallic catalysts for hydrogen production. *Catalysis Letters*, **101**, 355–64.

36 Salge, J.R., Deluga, G.A. and Schmidt, L.D. (2005) Catalytic partial oxidation of ethanol over noble metal catalysts. *Journal of Catalysis*, **235**, 69–78.

37 Morgenstern, D.A. and Fornango, J.P. (2005) Low-temperature reforming of ethanol over copper-plated Raney nickel: a new route to sustainable hydrogen for transportation. *Energy Fuels*, **19**, 1708–16.

38 Diagne, C., Idriss, H. and Kiennemann, A. (2002) Hydrogen production by ethanol reforming over Rh/CeO2-ZrO2 catalysts. *Catalysis Communications*, **3**, 565–71.

39 Roh, H.-S., Wang, Y., King, D.L., Platon, A. and Chin, Y.-H. (2006) Low temperature and H2 selective catalysts for ethanol steam reforming. *Catalysis Letters*, **108**, 15–19.

40 Avci, A., Trimm, D.L., Aksoylu, A.E. and Onsan, Z.I. (2004) Hydrogen production by steam reforming of n-butane over supported Ni and Pt-Ni catalysts. *Applied Catalysis A: General*, **258**, 235–340.

41 Wang, X. and Gorte, R.J. (2001) Steam reforming of n-Butane on Pd/Ceria. *Catalysis Letters*, **73**, 15–19.

42 Caglayan, B.S., Onsan, Z.I. and Aksoylu, A.E. (2005) Production of hydrogen over bimetallic Pt-Ni/d-Al2O3: II: indirect partial oxidation of LPG. *Catalysis Letters*, **102**, 63–7.

43 Ma, L., Trimm, D.L. and Jiang, C. (1996) The design and testing of an autothermal reactor for the conversion of light hydrocarbons to hydrogen I. The kinetics of the catalytic oxidation of light hydrocarbons. *Applied Catalysis A: General*, **138**, 275–83.

44 Zhang, J., Wang, Y., Ma, R. and Wu, D. (2003) Characterization of alumina-supported Ni and Ni-Pd catalysts for partial oxidation and steam reforming of hydrocarbons. *Applied Catalysis A: General*, **243**, 251–9.

45 Qi, A., Wang, S., Ni, C. and Wu, D. (2007) Autothermal reforming of gasoline on Rh-based monolithic catalysts. *International Journal of Hydrogen Energy*, **32**, 981–91.

46 Williams, K. and Schmidt, L.D. (2006) Catalytic autoignition of higher alkane partial oxidation on Rh-coated foams. *Applied Catalysis A: General*, **299**, 30–45.

47 Wang, L., Murata, K. and Inaba, M. (2005) Steam reforming of gasoline promoted by partial oxidation reaction on novel bimetallic Ni-based catalysts to generate hydrogen for fuel cell-powered automobile applications. *Journal of Power Sources*, **145**, 707–11.

48 Murata, K., Wang, L., Saito, M., Inaba, M., Takahera, I. and Mimura, N. (2004) Hydrogen production from steam reforming of hydrocarbons over alkaline-earth metal-modified Fe- or Ni-based catalysts. *Energy and Fuels*, **18**, 122–6.

49 Cheekatamarla, P.K. and Lane, A.M. (2005) Catalytic autothermal reforming of diesel fuel for hydrogen generation in fuel cells I. activity tests and sulfur poisoning. *Journal of Power Sources*, **152**, 256–63.

50 Binwale, R.B. and Ichikawa, N.K.M. (2005) Production of hydrogen-rich gas via reforming of Iso-Octane over Ni-Mn and Rh-Ce bimetallic catalysts using spray pulsed reactor. *Catalysis Letters*, **100**, 16–25.

51 Erri, P., Dinka, P. and Varma, A. (2006) Novel perovskite-based catalysts for autothermal JP-8 fuel reforming. *Chemical Engineering Science*, **61**, 5328–33.

52 Goud, S.K., Whittenberger, W.A., Chattopadhyay, S. and Abraham, M.A. (2007) Steam reforming of n-hexadecane using a Pd/ZrO2 catalyst: kinetics of catalyst deactivation. *International Journal of Hydrogen Energy*, **32**, 2868–74.

53 Cheekatamarla, P.K. and Thomson, W.J. (2006) Catalytic activity of molybdenum carbide for hydrogen generation via diesel reforming. *Journal of Power Sources*, **158**, 477–84.

54 Ouyang, X., Bednarova, L., Besser, R.S. and Ho, P. (2005) Preferential oxidation (PrOx) in a thin-film catalytic microreactor: advantages and limitations. *AIChE Journal*, **51**, 1758–72.

55 Dagle, R.A., Wang, Y., Xia, G.-G., Strohm, J.J., Holladay, J. and Palo, D.R. (2007) Selective CO methanaton catalysts for fuel processing applications. *Applied Catalysis A: General*, **326**, 213–18.

56 Franz, A.J., Jensen, K.F. and Schmidt, M.A. (1999) Palladium based micromembranes for hydrogen separation and hydrogenation/dehydrogenation reactions. *Proceedings of the IEEE: Microelectromechanical Systems (MEMS)*, 382.

57 Tong, H.D., Berenschot, E., De Boer, M.J., Gardeniers, J.G.E., Wensink, H., Jansen, H.V., Nijdam, W., Elwenspoek, M.C., Gielens, F.C. and van Rijn, C.J.M. (2003) Microfabrication of palladium-silver alloy membranes for hydrogen separation. *Journal of Microelectromechanical Systems*, **12**, 622–9.

58 Wilhite, B.A., Schmidt, M.A. and Jensen, K.F. (2004) Palladium-based micromembranes for hydrogen separation: device performance and chemical stability. *Industrial and Engineering Chemistry Research*, **43**, 7083–91.

59 Hamakawa, S., Li, L., Li, A. and Iglesia, E. (2002) Synthesis and hydrogen permeation properties of membranes based on dense SrCe0.95Yb0.05O3-a thin films. *Solid State Ionics, Diffusion and Reactions*, **148**, 71–81.

60 Qi, X. and Lin, Y.S. (1999) Electrical conducting properties of proton-conducting terbium-doped strontium cerate membrane. *Solid State Ionics*, **120**, 85–93.

61 Lundberg, W.L. and Veyo, S.E. (2003) Solid oxide fuel cell hybrid power system using the mercury 50 advanced turbine systems gas turbine. *Journal of Engineering for Gas Turbines and Power*, **125**, 51–8.

62 Mohamad, A.A., Ramadhyani, S. and Viskanta, R. (1994) Modeling of combustion and heat transfer in a packed bed with embedded coolant tubes. *International Journal of Heat and Mass Transfer*, **37**, 1181–91.

63 Viskanta, R. (1991) Enhancement of heat transfer in industrial combustion systems. Problems and future challenges. *ASME/JSME Thermal Engineering Joint Conference*, 161–73.

64 Levenspiel, O. (2005) What will come after petroleum? *Industrial and Engineering Chemistry Research*, **44**, 5073–8.

65 Ramaswamy, R.C., Ramachandran, P.A. and Dudukovic, M.P. (2006) Recuperative coupling of exothermic and endothermic reactions. *Chemical Engineering Science*, **61**, 459–72.

66 Douglas, J.M. (1988) *Conceptual Design of Chemical Processes*, McGraw-Hill, New York, NY.

67 Ehrfeld, W., Hessel, V. and Lowe, H. (2000) *Microreactors: New Technology for Modern Chemistry*, John Wiley & Sons, Ltd, Weinheim.

68 www.ehrfeld.com (12 October, 2008).

69 Keoschkerjan, R., Richter, M., Boskovic, D., Schnurer, F. and Lobbecke, S. (2004) Novel multifunctional microreaction unit for chemical engineering. *Chemistry-Engineering Journal*, **101**, 469–75.

70 Lobbecke, S., Wolfgang, F., Panic, S. and Tobias, T. (2005) Concepts for modularization and automation of microreaction technology. *Chemical*

Engineering and Technology, **28**, 484–93.

71 Palo, D.R., Holladay, J.D., Rozmiarek, R.T., Guzman-Leong, C.E., Wang, Y., Hu, J., Chin, Y.-H., Dagle, R.A. and Baker, E.G. (2002) Development of a soldier-portable fuel cell power system. Part I: a bread-board methanol fuel processor. *Journal of Power Sources*, **108**, 28–34.

72 Palo, D.R., Holladay, J.D., Rozmiarek, R.T., Guzman-Leong, C.E., Wang, Y., Hu, J. Chin, Y.-H., Dagle, R.A. and Baker, E.G. (2001) Fuel processor development for a soldier-portable fuel cell system, in *Microreaction Technology, IMRET 5 Proceedings of the 5th International Conference on Microreaction Technology* (eds Matlosz, M. Erfeld, W. and Baselt, J.P.), Springer, 359–67.

73 Peters, M.S. and Timmerhaus, K.D. (2003) *Plant Design and Economics for Chemical Engineers*, 5th edn, McGraw Hill, New York, NY.

74 Delsman, E.R., Kramer, M.H.J.M., de Croon, G.J., Cobden, P.D., Hofmann, C., Cominos, V. and Schouten, J.C. (2004) Experiments and modeling of an integrated preferential oxidation-heat exchanger microdevice. *Chemical Engineering Journal*, **101**, 123–31.

75 Besser, R. (2007) Small scale hydrogen production: status and future challenges, presented at ACS 231st Meeting, Boston, August 23rd, FUEL 248.

76 Arana, L.R., Schaevitz, S.B., Franz, A.J., Schmidt, M.A. and Jensen, K.F. (2003) A microfabricated suspended-tube chemical reactor for thermally efficient fuel processing. *Journal of Microelectromechanical Systems*, **12**, 600–12.

77 Schaevitz, S.B., Franz, A.J., Jensen, K.F. and Schmidt, M.A. (1994) A combustion-based MEMS thermoelectric power generator, Tranducers '01. Eurosensors XV. *11th International Conference on Solid-State Sensors and Actuators. Digest of Technical Papers*, **1**, 30–3.

78 Ducso, C., Vazsonyi, E., Adams, E.M., Szabo, I., Barsony, I., Gardeniers, J.G.E. and van den Berg, A. (1997) Porous silicon bulk micromachining for thermally isolated membrane formation. *Sensors and Actuators A (Physical)*, A**60**, 235–9.

79 Srinivasan, R., Hsing, I.-M., Berger, P.E., Jensen, K.F., Firebaugh, S.L., Schmidt, M.A., Harold, M.P., Lerou, J.J. and Ryley, J.F. (1997) Micromachined reactors for catalytic partial oxidation reactions. *AIChE Journal*, **43**, 3059–69.

80 Norton, D.G., Wetzel, E.D. and Vlachos, D.G. (2006) Thermal management in catalytic microreactors. *Industrial and Engineering Chemistry Research*, **45**, 76–84.

81 Federici, J.A., Norton, D.G., Bruggemann, T., Voit, K.W., Wetzel, E.D. and Vlachos, D.G. (2006) Catalytic microcombustors with integrated thermoelectric elements for portable power production. *Journal of Power Sources*, **161**, 1469–78.

82 Srinivasan, R., Hsing, I.-M., Ryley, J., Harold, M.P., Jensen, K.F. and Schmidt, M.A. (1996) Micromachined chemical reactors for surface catalyzed reactions, in *Solid-State Sensor and Actuator Workshop Proceedings*, Hilton Head, SC.

83 Yamamoto, N. (2006) Thermomechanical Properties and Performance of Microfabricated Solid Oxide Fuel Cell (mSOFC) Structures. M.S. Thesis, Massachusetts Institute of Technology.

84 Jankowski, A.F., Hayes, J.P., Graff, R.T. and Morse, J.D. (2002) Microfabricated thin-film fuel cells for portable power requirements. *Materials for Energy Storage, Generation and Transport Symposium, Materials RESEARCH Society Proceedings*, **730**, 93–8.

85 Shah, K., Shin, W.C. and Besser, R.S. (2004) A PDMS micro proton exchange membrane fuel cell by conventional and non-conventional microfabrication techniques. *Sensors and Actuators B*, **97**, 157–67.

86 Maynard, H.L. and Meyers, J.P. (2002) Miniature fuel cells for portable power: design considerations and challenges. *Journal of Vacuum Science and Technology, B.*, **20**, 1287–93.

87 Hayase, M., Kawase, T. and Hatsuzawa, T. (2004) Miniature 250 mm thick fuel cell with monolithically fabricated silicon electrodes. *Electrochemical and Solid-State Letters*, **7**, A231–234.

88 Mukherjee, S., Hatalis, M.K. and Kothare, M.V. (2007) Water gas shift reaction in a glass microreactor. *Catalysis Today*, **120**, 107–20.

89 London, A.P., Ayon, A.A., Epstein, A. H., Spearing, S.M., Harrison, T., Peles, Y. and Kerrebrock, J.L. (2001) Microfabrication of a high pressure bipropellant rocket engine. *Sensors and Actuators, A: Physical*, **92**, 351–7.

90 Mirza, A.R. and Ayon, A.A. (1999) Silicon wafer bonding for MEMS manufacturing. *Solid State Technology*, **42**, 73–8.

91 de Mas, A., Gunther, T., Kraus, M.A., Schmidt, K.F. and Jensen, K.F. (2005) Scaled-out multilayer gas-liquid microreactor with integrated velocimetry sensors. *Industrial and Engineering Chemistry Research*, **44**, 8997–9013.

92 Inoue, T., Schmidt, M.A. and Jensen, K. F. (2007) Microfabricated multiphase reactors for the direct synthesis of hydrogen peroxide from hydrogen and oxygen. *Industrial and Engineering Chemistry Research*, **46**, 1153–60.

93 Livermore, C., Hill, T.F., Velaquez-Garcia, L., Wilhite, B.A., Epstein, A.H., Jensen, K.F., Rawlins, W.T., Lee, S. and Davis, S. (2007) Singlet-oxygen generator on a chip for MEMS-based COIL. *Proceedings of SPIE – The International Society for Optical Engineering*, **6454**, 64540G.

94 Sahoo, H., Kralj, J.G. and Jensen, K.F. (2007) Multistep continuous-flow microchemical synthesis involving multiple reactions and separations. *Angewandte Chemie – International Edition*, **46**, 5704–8.

95 Ajmera, S.K., Delattre, C., Schmidt, M. A. and Jensen, K.F. (2002) Microfabricated cross-flow chemical reactor for catalyst testing. *Sensors and Actuators, B: Chemical*, **82**, 297–306.

96 Besser, R.S., Ouyang, X., Surangalikar, H. and Prevot, M. (2001) Microreactor for efficient catalyst evaluation. *Proceedings of the SPIE – The International Society for Optical Engineering*, **4560**, 75–82.

97 Easley, C.J., Humphrey, J.A.C. and Landers, J.P. (2007) Thermal isolation of microchip reaction chambers for rapid non-contact DNA amplification. *Journal of Micromechanics and Microengineering*, **17**, 1758–66.

98 Reuse, P., Renken, A., Haas-Santo, K., Gorke, O. and Schubert, K. (2004) Hydrogen Production for Fuel Cell Application in an Autothermal Micro-Channel Reactor, *Chemistry-Engineering Journal*, **101**, 133–41.

99 Irving, P.M., Allen, W.L., Healey, T., Ming, Q. and Thomson, W.J. (2002) Catalytic micro-reactor systems for hydrogen generation, in Microreaction Technology, IMRET 5 Proceedings of the 5th International Conference on Microreaction Technology (eds M. Matlosz, W. Erfeld and J.P. Baselt), pp. 286–94.

100 Walter, S., Joannet, E., Schiel, M., Boullet, I., Phillips, R. and Liauw, M.A. (2002) Microchannel reactor for the partial oxidation of isoprene, in Microreaction Technology, IMRET 5 Proceedings of the 5th International Conference on Microreaction Technology (eds M. Matlosz, W. Erfeld and J.P. Baselt), pp. 387–96.

101 Mies, M.J.M., Rebrov, E.V., de Croon, M. H.J.M. and Schouten, J.C. (2004) Design of a molybdenum high throughput microreactor for high temperature screening of catalytic coatings. *Chemistry-Engineering Journal*, **101**, 225–35.

102 Rouge, A., Spoetzl, B., Gebauer, K., Schenk, R. and Renken, A. (2001) Microchannel reactors for fast periodic operation: the catalytic dehydrogenation of isopropanol. *Chemical Engineering Science*, **56**, 1419.

103 Knitter, R., Gohring, D., Risthaus, P. and Haubelt, J. (2001) Microfabrication of ceramic microreactors. *Microsystem Technologies*, **7**, 85–90.

104 Schmitt, C., Agar, D.W., Platte, F., Buijssen, S., Pawlowski, B. and Duisberg, M. (2005) Ceramic plate heat exchanger for heterogeneous gas-phase reactions. *Chemical Engineering & Technology*, **28**, 337–43.

105 Meschke, F., Riebler, G., Hessel, V., Schurer, J. and Baier, T. (2005) Hermetic gas-Tight ceramic microreactors. *Chemical Engineering & Technology*, **28**, 465–73.

106 Stief, T., Langer, O.-U. and Schubert, K. (1999) Numerical investigations of optimal heat conductivity in micro heat exchangers. *Chemical Engineering & Technology*, **21**, 297–303.

107 Peterson, R.B. (1999) Numerical modeling of conduction effects in microscale counterflow heat exchangers. *Microscale Thermophysical Engineering*, **3**, 17–30.

108 Wada, Y., Schmidt, M.A. and Jensen, K. F. (2006) Flow distribution and ozonolysis in gas-liquid multichannel microreactors. *Industrial and Engineering Chemistry Research*, **45**, 8036–42.

109 Pan, T., Baldi, A. and Ziaie, B. (2006) A reworkable adhesive-free interconnection technology for microfluidic systems. *Journal of Microelectromechanical Systems*, **15**, 267–72.

110 Ratner, D.M., Murphy, E.R., Jhunjhunwala, M., Snyder, D.A., Jensen, K.F. and Seeberger, P.H. (2005) Microreactor-based reaction optimization in organic chemistry-glycosylation as a challenge. *Chemical Communications*, **5**, 578–83.

111 Inoue, T., Murphy, E.R., Schmidt, M.A. and Jensen, K.F. (2003) Microreactor direct synthesis of hydrogen peroxide from hydrogen and oxygen, in 7th international conference on microreaction technology (IMRET 7), Lausanne, Switzerland.

112 Peles, Y., Srikar, V.T., Harrison, T.S., Protz, C., Mracek, A. and Spearing, S.M. (2004) Fluidic packaging of microengine and microrocket devices for high-pressure and high-temperature operation. *Journal of Microelectromechanical Systems*, **13**, 31–40.

113 Gonzalez, C., Collins, S.D. and Smith, R.L. (1998) Fluidic interconnects for modular assembly of chemical microsystems. *Sensors and Actuators B-Chemical*, **49**, 40–5.

114 Deshpande, K.T., Wilhite, B.A., Schmidt, M.A. and Jensen, K.F. (2005) Integrated partial oxidation and purification microsystems for autothermal production of hydrogen from methanol. AIChE Annual Meeting and Fall Showcase, Conference Proceedings, pp. 9433–4.

115 Meng, E., Wu, S., Tai, Y.-C. (2001) Silicon couplers for microfluidic applications, *Fresenius' Journal of Analytical Chemistry*, **371**, 270–5.

116 Gray, B.L., Jaeggi, D., Mourlas, N.J., van Drieenhuizen, B.P., Williams, K.R., Maluf, N.I. and Kovacs, G.T.A. (1999) Novel interconnection technologies for integrated microfluidic systems, *Sensors and Actuators*, **77**, 57–65.

117 Murphy, E.R. (2006) Microchemical Systems for Rapid Optimization of Organic Syntheses, Ph.D. Thesis, Massachusetts Institute of Technology.

Part Two   System Design

Microfabricated Power Generation Devices. Edited by Alexander Mitsos and Paul I. Barton
Copyright © 2009 WILEY-VCH Verlag GmbH & Co. KGaA, Weinheim
ISBN: 978-3-527-32081-3

8
Selection of Alternatives and Process Design

Alexander Mitsos and Paul I. Barton

8.1

Introduction

The focus of this chapter is the comparison of various alternatives at the system level, for example is a device based on a proton exchange membrane (PEM) fuel cell advantageous over a micro-engine, and the choice of an optimal alternative. This decision resembles macroscale process synthesis, which is a very mature field, see for example [1]. There are however, some very important differences compared to macroscale process synthesis:

1. The components of a microfabricated device are highly coupled and cannot be seen in the unit operations design paradigm. Layout needs to be considered simultaneously with process synthesis [2].

2. There is a strong interaction between design and operation [3] and presumably between operation and control.

3. The objectives and constraints for the design of micropower generation devices are very different to those for macroscale processes. In the macroscale system efficiency and cost are the dominant criteria, while in power generation devices maximal energy (and/or power) density is required.

4. The large number of process alternatives for man-portable power generation arises from the large choice of fuels, fuel reforming reactions and fuel cells and the early stage of technology development, rather than from an elaborate combination of mixing, reaction and separation steps.

5. The processes must operate fully autonomously, automatically and without any safety concern, such as the use or generation of toxic or dangerous materials. It is paramount in computing energy density to have a process that operates independently of external heat sources despite, for example, the use of endothermic fuel processing reactions.

Microfabricated Power Generation Devices. Edited by Alexander Mitsos and Paul I. Barton
Copyright © 2009 WILEY-VCH Verlag GmbH & Co. KGaA, Weinheim
ISBN: 978-3-527-32081-3

In the remainder of this chapter, first the design objectives and constraints are analyzed., then potential methodologies for the comparison of alternatives are discussed and the state-of-the-art is summarized. Finally, results from illustrative case studies are presented.

8.2
Objectives, Constraints and Alternatives

8.2.1
Analysis of Design Objectives

In large-scale power production, emphasis is placed on efficient utilization of the fuel. This is partly because the life cycle fuel cost is of the same or higher order of magnitude as the fabrication cost of the power production system. In man-portable power production the economic and ecological operating costs are much smaller relative to the fabrication costs of the systems. Moreover, there are several applications for man-portable applications where performance outweighs cost considerations. Therefore, different metrics are needed for the design of micropower generation devices.

Typically, different man-portable power generation systems are compared using the metric of energy density of the system [4] (see below for a detailed analysis of this objective). Another important metric is the power density (power produced per volume/mass). Other quantifiable metrics include undesired heat signature for the dismounted soldier, undesired heat generation for personal devices and life span of a device. Depending on the application, different objectives are more important, which mandates product engineering, that is different designs for different end-users [2]; the power generation device for "One Laptop per Child" [5] will be very different to the power generation device for a dismounted soldier. In some applications economic and environmental calculations are important, at least as a secondary objective. For instance, it would be interesting to minimize the cost of the device subject to an acceptable decrease in performance over the optimal design. Also, the influence of multiple objectives is interesting; for instance in [6] the trade-off between two objectives is demonstrated by calculating the corresponding Pareto curve.

8.2.1.1 Efficiency vs. Energy Density
A common misconception is that efficiency of power generation devices must be maximized. In the following it is analyzed that maximal efficiency does not necessarily imply maximal energy density. The specific energy, or *gravimetric energy density* e_{grav}^{sys} [W h kg^{-1}], is expressed as the electrical energy produced per unit mass of system [7] and the *volumetric energy density* e_{vol}^{sys} [W h l^{-1}] is defined as the electrical energy produced per unit volume of the system

$$e_{grav}^{sys} = \frac{\tau_{mission} PW}{M^{sys}}, \quad e_{vol}^{sys} = \frac{\tau_{mission} PW}{V^{sys}},$$

where the mission duration τ_{mission} [h] is the time between refueling or recharging, PW [W] is the power output (assumed constant for simplicity), M^{sys} [kg] is the mass of the system (including fuel and oxidant if carried), and V^{sys} [5] is the volume of the system. Depending on the application, either of the densities is more important. It is essential to define the system appropriately including the power generation devices as well as the fuel and oxidant containers. The most promising application of fuel cell systems is for long mission durations [2, 8]. In the limit of an infinite mission duration the system energy density approaches the *fuel energy density*, that is the power produced per unit mass flow or volumetric flow of the fuel and possibly oxidant

$$e_{\text{grav}}^{\text{fuel}} = \frac{\text{PW}}{3600 \sum_i \text{MW}_i N_{i,\text{in}}}, \quad e_{\text{vol}}^{\text{fuel}} = \frac{\text{PW}}{3600 \sum_i \text{MV}_i N_{i,\text{in}}},$$

where $N_{i,\text{in}}$ [mol s^{-1}] is the inlet molar flow rate of species i, MW_i [kg mol^{-1}] is the molecular weight of species i, MV_i [l mol^{-1}] is the molar volume of species i at storage conditions, 3600 is the conversion factor from hours to seconds, and the summation is taken over all stored fuels and oxidants.

Often it is important to define the energy conversion efficiency of the components of a system, for example the reforming reactor or the fuel cell, and sometimes alternative definitions of efficiency are plausible. For instance, the efficiency of the fuel cell can be defined as the quotient of power produced to chemical energy fed to the fuel cell per unit time or as the quotient of power produced to chemical energy consumed per unit time. One definition accounts for unreacted fuel while the other does not. The latter definition is more appropriate when the fuel cell effluents are utilized in a burner. As is shown in the following, for the case of one chemical reaction, the fuel cell voltage is an appropriate metric for the energy conversion efficiency. The definition of the component efficiency gives

$$\eta_{\text{FC}} = \frac{\text{PW}}{\xi \Delta_r H(T_{\text{amb}})},$$

where η_{FC} is the fuel cell efficiency, PW [W] is the power, ξ [mol s^{-1}] is the extent of the reaction, and $\Delta_r H(T_{\text{amb}})$ [J mol^{-1}] is the corresponding enthalpy of reaction at ambient conditions. The rationale for using ambient conditions is to have a common metric for processes working at different conditions. On the other hand, the power produced is given by the product of voltage U [V] and current I [A]

$$\text{PW} = UI$$

The current is the product of the extent of reaction, the Faraday constant F [C mol^{-1}], and the number of electrons z exchanged in the oxidation of hydrogen (2 per molecule of H_2)

$$I = z\xi F$$

and therefore

$$\eta_{FC} = u \frac{Fz}{\Delta_r H(T_{amb})} .$$

The overall energy conversion efficiency η^{sys} for a power system is given by the power generated divided by the chemical energy fed to the system per unit time, accounting for unreacted fuel

$$\eta^{sys} = \frac{PW}{\sum_i H_i^{heat} N_{i,in}} \qquad (8.1)$$

where H_i^{heat} [J mol^{-1}] is the heating value of species i, that is the heat released from complete combustion of one mole of species i at reference conditions $T_{ref} = 298$ K. Note that here the energy stored in the oxidant (e.g. the compression energy) is ignored. The system energy conversion efficiency is a function of the component efficiencies. For components in series the overall efficiency is the product of the individual efficiencies; for components in parallel the overall efficiency is the average of the component efficiencies weighted with the individual power levels. Improving an individual component efficiency does not necessarily imply that the energy conversion efficiency of the system will also be improved [6]. This is due to the often complicated interaction between individual components.

The objective of maximal energy density is, in general, not equivalent to the objective of maximal efficiency. Energy efficiency is not a suitable metric for selecting, designing or operating portable power generation devices, that is, neither for choosing a given process among a list of candidate alternatives nor for finding the optimal design and operation parameters of a given alternative [6]. The simplest example illustrating this is a comparison between different fuels; choosing a fuel with high energy density can lead to a higher system energy density despite a lower efficiency. For instance a 35% efficient butane system has a higher energy density than a 70% efficient ammonia system (see the figure in the Introductory chapter). Similar behavior is seen for systems that use a combination of different fuels/ chemicals, where energy density and energy conversion efficiency bear different weights on each species. The extreme case of species combination is the addition of water in steam reforming reactions, which does not (directly) affect the energy efficiency but greatly reduces the energy density, assuming that the water cannot be recycled but has to be carried along with the fuel. Another extreme example is the case where no ambient air is available for the oxidation reactions, for example for underwater or space applications, and an oxygen or air cartridge must be carried. The gases require significant storage volumes and cartridge mass and the energy density is significantly lower than with the use of ambient air [9]. On the other hand, the energy stored in the compressed gas is very small compared to the chemical energy of the fuels and, therefore, the metric of energy conversion efficiency does not account for the stored gas. Moreover, the use of pure oxygen may increase the energy conversion efficiency compared to the use of ambient air. For systems involving only one stored species i, the fuel energy density and efficiency are proportional, since

$$e_{grav}^{fuel} = \frac{PW}{3600 MW_i N_{i,in}}, \quad e_{vol}^{fuel} = \frac{PW}{3600 MV_i N_{i,in}}, \quad \eta^{sys} = \frac{PW}{H_i^{heat} N_{i,in}}$$

and therefore

$$\frac{e_{grav}^{fuel}}{\eta^{sys}} = \frac{H_i^{heat}}{3600 MW_i}, \quad \frac{e_{vol}^{fuel}}{\eta^{sys}} = \frac{H_i^{heat}}{3600 MV_i}$$

On the other hand the system energy density is also a function of the device size, while the energy conversion efficiency is not.

8.2.1.2 Safety Concerns, High Temperatures and Heat Dissipation

A common criticism of micropower generation devices, and particularly of high temperature systems, is that they pose safety threats and generate a lot of heat. In the following it is argued that these concerns are partially true and partially misconceptions resulting from macro-scale experience. Direct contact with the hot surface of the reactors or fuel cell would cause harm, but the insulation acts as a protection.

Recall that the promise of micropower generation devices is a higher energy density than state-of-the-art batteries, by using fuels with a high intrinsic energy density. As a direct consequence, safety concerns (e.g. explosion after a thermal/ mechanical shock) arise with the stored fuel and, therefore, innovative storage options need to be considered, see for example [10]. In contrast, the power generation device itself does not pose such a serious threat, as suggested by the following comparison of stored energy. Consider a power generation device for $PW = 10\,W$ operating at $T = 1000\,K$. Suppose that the volume of the hot box is $V = 1\,cm^3$, the gas residence time in the device $\tau = 10\,s$ and the thermal efficiency $\eta = 0.1$. Taking the worse case scenario that the box is solid silicon, the thermal energy stored is only $\approx 1\,kJ$. The chemical energy of the gases in the device can be estimated by the power produced times the residence time divided by the thermal efficiency, also $\approx 1\,kJ$. Similarly, the chemical energy of the gases in the storage tank can be estimated by the power produced times the mission duration time divided by the thermal efficiency. For a mission duration of just 10h this gives 3.6 MJ. This simple calculation shows that the thermal energy stored is of the same order of magnitude as the chemical energy of the gases in the device, and that they are both much smaller than the energy stored in the fuel cartridge. In considering these concerns one should not forget that high-performance batteries can also explode, and safety standards have been established, see for example [11].

Another safety concern is the toxicity of the chemicals stored and produced. Despite the small amounts of fuel used, this is a concern that needs to be considered in the design [3]. Moreover, leaking components and breakage of the storage are important. Some of these concerns also apply to batteries, but are not as serious since most batteries are closed systems.

Power generation is associated with heat generation, inversely proportional to the overall system efficiency, see for example [12]. This holds for any power generation device, irrespective of the technology used, whether it is a battery, micro-fuel cell, micro-engine, or even a radioactive element. Inefficient processes might be considered uncomfortable for portable applications because of the large heat generation, for example a cellular phone getting hot, or yield an undesired heat signature in the battlefield. An inefficient low-temperature device will generate more heat than a highly efficient high-temperature device. This contradicts the perception that high temperature will lead to high heat losses. This perception likely originates from considering a fixed surface area and no insulation. In low-temperature processes, high thermal conductivity is desired and heat dissipation by conduction and natural convection may not be sufficient [12]. In contrast, for high-temperature systems to be viable, good heat insulation is required, which as a side-effect minimizes the undesired heat dissipation. Moreover, as discussed Chapter 11, dedicated to optimal design and operation, higher temperatures lead to faster reactions, therefore smaller residence time requirements, therefore lower surface area and, in some cases, overall lower heat losses [3].

8.2.2
Decisions

In designing a microfabricated power generation device one must decide on many features of the device. The first design choice is the fuel(s) that will be used for power and heat production. A basic requirement for a fuel is that it is compressible at relatively low pressures, so that it occupies a small volume. Ideally, the fuels used should be nontoxic and inherently safe, but this requirement can be relaxed by allowing fuels that pose health and safety concerns similar to chemicals used in common consumer applications. A variety of fuels are being considered by different research groups, including ammonia, methanol/ethanol mixtures, propane/butane mixtures, methane, compressed hydrogen, hydrides (hydrogen generators) and formic acid. In terms of oxidants, atmospheric air, compressed oxygen, and compressed air, as well as oxygen generators (e.g. hydrogen peroxide), are considered. Among these chemicals some are very toxic, such as ammonia, and could only be used in remote applications. Others, such as methane, cannot be compressed adequately and their use is very unlikely for real-world applications.

The second design choice is whether to perform fuel processing in a reactor or to feed the fuel to a fuel cell directly. In earlier chapters the various fuel processing alternatives and fuel cell technologies were discussed. From a process design perspective it is important to note that, based on the process design heuristic for simplicity [13], only very few "units" are likely to be included in micropower devices.

The next design choice is whether this fuel or a secondary fuel will be fed into a burner for heat generation. The heat produced from burners serves to compensate for stream preheating, heat losses, endothermic reactions or even heating of the system at start-up.

Certain chemical species, such as carbon monoxide, have deleterious effects on some fuel cells, for example PEM, and it may, therefore, be necessary to perform a gas purification, for instance using a palladium membrane for hydrogen separation. Such a membrane will incur partial loss of the hydrogen in the waste stream and a sweep stream may be necessary for operation [14]. The purification could either be sequential to the reactor, or the reactor and the membrane could be combined into one unit, shifting the equilibrium towards hydrogen [14, 15]. The separation waste can be either discarded or burned. If desired, the purification product (H_2) can be split, and a part can be fed into a burner.

A variety of fuel cells are being considered by research groups, including solid oxide fuel cells (SOFCs) (with or without internal reforming), proton ceramic fuel cells (PCFCs), hydrogen operated polymer electrolyte membrane fuel cells (PEM), single chamber fuel cells operating with hydrogen and carbon monoxide, and direct methanol fuel cells (DMFCs), or even biofuel cells [16]. A SOFC has the benefit of fuel flexibility, but it is operated at high temperature which leads to large heat losses and difficult start-up. PEMs are run at low temperatures but cannot tolerate impurities, and water management is an issue. Single chamber fuel cells are potentially easier to fabricate [17], but have the drawback that they are operated with premixed gases which, potentially, can lead to explosions and require catalysts with high selectivity. A PCFC is a relatively new concept [18], which has the potential of fuel flexibility while operating at slightly lower temperatures than SOFCs. A DMFC is a PEM-based fuel cell in which a dilute methanol solution in water is reformed at a relatively low temperature, around 350 K; major technical challenges include methanol crossover and water management. Recall that technological aspects of fuel cells are detailed in an earlier chapter.

The conversion in the fuel cells (also denoted "fuel utilization") is usually not complete, and the unreacted part of the fuel can either be burned or recycled. A more promising recycling option would be recycling after separation, for example separate the hydrogen of the fuel cell effluent and then recycle it to the fuel cell, or separate the steam/water and use it for reforming reactions and to prevent coking. These options are very appealing from the point of view of minimizing the mass of stored fuels but separation might be very difficult to implement in the general case. Depending on the implementation of the recycle stream, a pressure increase mechanism may be necessary, for example a microfabricated pump, which will result in an energetic penalty in terms of a compression power.

The cathode effluent stream of the fuel cell can be reused to provide oxygen for a burner because it is plausible that the fuel cells will be operated at a relatively large oxygen excess. Reusing excess oxygen is most advantageous in volume-critical applications, where the oxygen cartridge may occupy a large fraction of the total system volume. In addition, the temperature of the cathode effluent stream is higher than the ambient, so this reduces the energetic requirement of preheating the oxygen feed to the burner. However, in circumstances where the fuel cell discard temperature is substantially lower than the operating temperature of the burner (i.e. for a PEM or DMFC), preheating is still necessary. The cathode effluent also contains nitrogen and, in some cases, for example a PEM, also steam, and heating of these components to the burner operating temperature may outweigh the advantage of using preheated oxygen.

8.3
Methodologies

Macroscale process synthesis is a mature field and several approaches exist for the comparison of alternatives and conceptual process design. While on the macroscale there are few limitations for process synthesis, on the microscale only relatively simple processes are possible [13, 19]. However, the large choice of fuels, fuel reforming reactions and fuel cells results in many potential alternatives. Moreover, the early stage of technology development, along with the very different technologies available (e.g. microfabricated fuel cell vs. micro-engine), makes the comparison very challenging.

Since it is impossible to fabricate and test all alternatives, a model-based analysis seems necessary. To be independent of technological details, such as the catalysts used or the reactor configuration, general and, as a consequence, relatively simple models, must be used. This system-level analysis provides estimates of the size of the device and limits of performance and can be used to determine at an early stage if the development of a proposed device is worth pursuing. As an example, the use of methane, which has been proposed in the literature, is shown in the case studies to be marginally competitive with existing battery technologies, because of the storage requirements, even assuming high conversion efficiencies. Based on the macroscale process synthesis literature many methodologies could be used for this system-level analysis.

One of the possible methodologies for comparison of alternatives is the notion of a superstructure. A superstructure contains all the alternatives to be considered in the selection of an optimal process structure [1]. An actual process design is a subset of the units and connections in the superstructure. In [2, 9] a superstructure approach was used for microfabricated fuel cell systems. Through the use of simulation and (parametric) mixed-integer optimization the most promising process structures together with idealized layouts are selected from among thousands of alternatives. This approach is described in more detail below. A similar concept to superstructures is exergy analysis (or availability analysis). This approach calculates limits on process performance based on thermodynamic limits [20]. In the context of micropower generation it has been used in [21, 22], wherein alternatives based on methanol, ethanol, octane, ammonia, and methane are considered for the generation of 100 W of electricity. There are several other approaches in macroscale process synthesis which, to the best knowledge of the authors, have not yet found application to micropower or micro-chemical systems. One of these alternatives is phenomena-based process synthesis [23]. Instead of considering unit operations (e.g. reactor, decanter) the phenomena-based approach builds a process from a combination of the required physical phenomena (e.g. reaction, separation). This abstraction can lead to new process combinations or combined units (e.g. reactive distillation). Another concept is that of attainable regions [24–26], where the full set of product composition vectors that can be produced (at least in principle) by reaction and mixing is considered. Also, the state-space approach, for example [27], is used in macroscale process synthesis, where the behavior of a system is described quantitatively in terms of a set of variables and a set of relations among these variables.

8.3.1
A Superstructure-Based Approach

In [2, 9, 28] alternative microfabricated fuel cell systems, chosen with the constraint that the realization of the processes is either currently under investigation or foreseeable in the short-term future (coming years) were represented with a conceptual superstructure, Figure 8.1. The superstructure contains design choices which, in Figure 8.1, are represented with hexagons and mathematically are represented with binary (0–1) variables. The symbols used in Figure 8.1 are explained in Figure 8.2. The various units (reactor, burners, etc.) should not be interpreted in the traditional unit operation design paradigm, but rather as closely interconnected parts of an integrated process. The superstructure is only conceptual and does not include any information about the physical layout of the units.

The set of alternatives is represented mathematically as a set of nonlinear algebraic equations (steady-state model). These system-level models account for thermal integration of the processes, as described in detail in the following section. The models are based on user-specified efficiency parameters in the various units, such as conversion, electrochemical efficiency, separation efficiency, and so on. Once these parameters, as well as the operating conditions, have been specified, the performance of the system is calculated.

Technological constraints do not allow for all potential process structures, for example carbon monoxide is a poison for PEM and therefore butane partial oxidation can only be combined with a PEM after gas purification. The optimal process structure depends on technological advances and product specifications. Similar to macroscale process synthesis, comparison of alternatives can be done through the specification of degrees of freedom and simulation of the resulting options, or for the automatic identification of optimal structures through mixed-integer optimization. The effect of technological parameters, such as the achievable conversion, can be studied by local sensitivity analysis, parametric studies or, ideally, parametric optimization [29].

8.3.2
Integrated Layout and Thermal Management

The graphical representation of the superstructure (Figure 8.1) does not contain information about the physical layout. In [9] the pronounced effect of heat losses on process performance and the importance of thermal management for high-temperature micropower generation devices were demonstrated. As described in Chapter 7 dedicated to thermal management, a very promising approach for thermal management is to couple two or more units thermally in a near-isothermal stack [30]. In this manner direct heat transfer between heat sinks and heat sources is possible, as well as heat recovery of the effluent streams; thermally

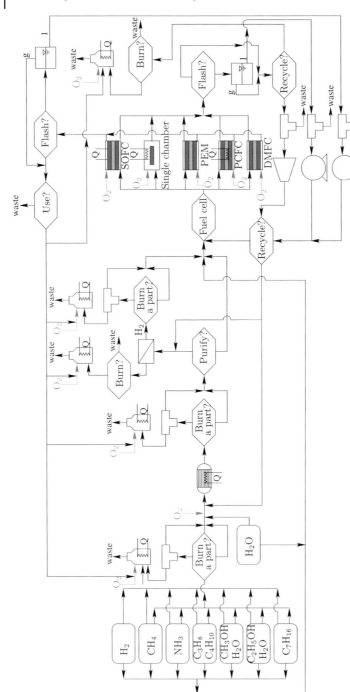

Figure 8.1 Set of alternatives considered.

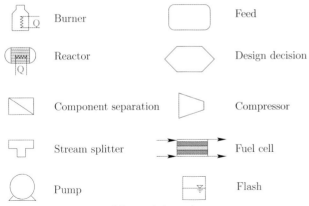

Figure 8.2 Explanation of the symbols used.

coupling two units also reduces the surface area and, as a consequence, the heat losses. Combining units is thus a layout consideration that influences the process performance. As a consequence, the problems of flow-sheet design, physical layout and heat integration need to be considered simultaneously.

Specification of the process layout requires an indication of the relative location and connectivity of every unit present in the selected flow-sheet of the superstructure, for example is the fuel cell in thermal contact with the reactor? In [2] an idealization of the layout considerations was proposed, allowing only for two extremes, implemented using logical decisions for the connectivity of each pair of units. One extreme is that the units are thermally connected, so that they share the surface through which heat losses occur and one energy balance is sufficient. Within these "stacks", flow streams proceed directly between units, so that heat losses between units are neglected. Due to the small length scales associated with these microdevices, convection and conduction are rapid in these stacks, thus necessitating that all units within a given stack operate with small temperature differences between them. The combination of units is limited by fabrication possibilities. The other extreme considered is that the units are spatially separated, so that each one has separate heat losses to the ambient and significant heat losses occur when mass and/or heat are transported between the units. Possible realizations of heat exchange are radiation or conduction through a rod connecting the units, or preheating of inlet gases to one unit by the heat excess of the other.

In Figure 8.3 the two extreme cases of layout are illustrated for a process in which butane is partially oxidized and the syngas produced is fed into a SOFC, while the cathode effluents are used to oxidize the unconverted syngas from the anode for heat generation. In one extreme case the SOFC, reactor and burner are assumed to be thermally coupled, while in the other extreme case all three process components are separate, with remote heat exchange between the burner and the SOFC. It should be noted that Figure 8.3 is conceptual and not an actual design. Similarly to the combination of units, fuel cartridges can be combined. This may

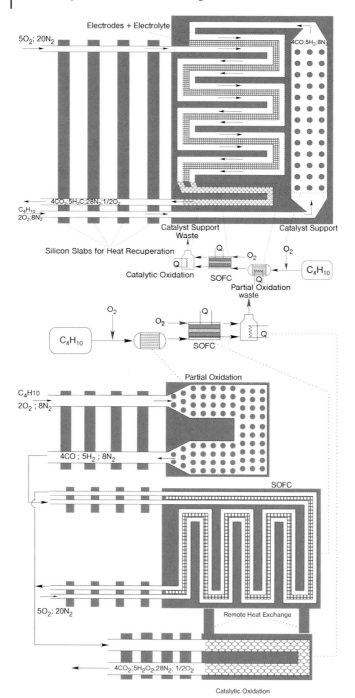

Figure 8.3 Conceptual difference between coupled (left) and noncoupled (right) process components.

be more difficult to fabricate, but has the advantage of minimizing the cartridge volume.

8.4
Case Studies

In this section some case studies are presented to demonstrate the value of modeling at the system level. Due to the early stage of development the parameter values used are not exact, but rather best available estimates. The numerical results depend on the values of the operating and technological parameters used. However, most qualitative results given are expected to be valid for the expected range of parameter values.

8.4.1
Comparison of Alternatives in Terms of Efficiency and Energy Density

Out of the thousands of possible process configurations and layouts in Figure 8.1, here four alternatives are compared in terms of the achievable energy densities and energy efficiency. For the model details, including the calculation of fuel and system energy density the reader is referred to [2, 9].

For all four alternatives a power generation of $PW = 10\,W$ is considered. For the calculation of system energy density a mission duration of $\tau_{mission} = 30\,h$ is assumed. Heat losses are accounted for with an overall transfer coefficient $U_{loss} = 3\,W\,m^{-2}\,K^{-1}$ and an emissivity (including the view factor) of $\varepsilon = 0.2$. The operating pressure is assumed to be 1 atm for all four processes. Three out of the four processes employ atmospheric air. For these processes a pressure increase mechanism such as a microblower is assumed, with a power requirement of 3 kJ per mol of air fed to the system. This conservative power requirement is calculated assuming isothermal compression at ambient temperature from 1 to 1.2 atm with an efficiency of 15% [9]. The SOFC and burner operating temperatures are set to 1000 K and the outlet temperatures to 600 K. The conversion is assumed to be 90% in the reactor, 80% in the fuel cell and 95% in the burner. The residence time in the SOFC is assumed to be 20 ms and in the burner 1 ms. The effect of packaging and thermal insulation are accounted for by calculating the device volume as 10 times the inner volume of the reactors and fuel cell. A device density of $1\,kg\,l^{-1}$ is assumed. The fuel cell power output is assumed to be 70% of the product of the extent of reaction and the standard Gibbs energy of reaction calculated at the fuel cell temperature

$$PW = 0.7\sum_r \xi_r \sum_r v_{ri} G_i^0(T)$$

where ξ_r is the extent of electrochemical reaction r, v_{ri} is the stoichiometric coefficient of species i in reaction r, and $G_i^0(T)$ is the molar gas phase Gibbs

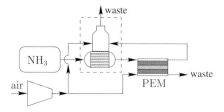

Figure 8.4 Flow-sheet for ammonia cracking.

energy of pure species *i* at the reference pressure. This simplification eliminates the dependence on composition.

The first alternative considered is the combination of ammonia cracking for the generation of hydrogen with a polymer electrolyte membrane (PEM) fuel cell, see Figure 8.4. While ammonia is very toxic, it is considered as a hydrogen source, see for example [31], because the cracking products do not contain carbon monoxide, which is a poison for PEM fuel cells. To overcome heat losses and the endothermicity of ammonia decomposition, the fuel cell effluent from the anode compartment is oxidized. Note that using a fresh air stream for the hydrogen burner leads to a better performance than using the air excess from the fuel cell effluent of the cathode compartment, because of the low temperature of the PEM and the nitrogen and water that are present in it. The hydrogen burner and the cracking reactor are considered to be in an isothermal stack for better performance [2]. The residence time in the reactor is assumed to be 1 ms, the operating temperature 1000 K and the outlet temperature 600 K, accounting for partial heat recovery. Atmospheric air is used for the oxidation reactions with an energetic penalty as described above. The equilibrium concentration of ammonia at 1000 K is approximately 200 ppm and therefore a significant performance decrease of the PEM may be observed [32, 33]. The models used do not account for this degradation but for actual deployment this issue has to be addressed. A possibility is to use an ammonia sorbent, see for example [34]. Also the need for cooling or humidifying of the PEM is neglected.

The second alternative considered is methane oxidation in a direct SOFC using compressed oxygen for the oxidation, see Figure 8.5. At the macroscale, steam reforming of natural gas is the predominant technology for hydrogen production. Methane as a feedstock for hydrogen production has the advantage that no carbon bonds need to be broken and the ratio of hydrogen-to-carbon atoms is maximal among the hydrocarbons. Moreover, compared with methanol and formic acid, methane has the advantage that it is not already partially oxidized. As a feedstock for portable applications it has the major drawback that it is supercritical at ambient temperatures and most likely to be stored as a compressed gas, which

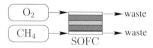

Figure 8.5 Flow-sheet for methane oxidation in a SOFC.

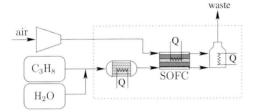

Figure 8.6 Flow-sheet for steam reforming of propane.

leads to low energy densities. The ideal case of direct electrochemical oxidation of methane in a SOFC is considered. As a further idealization the effect of carbon deposition on the catalysts is ignored. Gas storage is assumed in a plastic container with a maximal stress σ_{max} = 100 MPa, a density of ρ = 1.5 kg m^{-3} and a storage pressure of 10 MPa.

The third alternative considered is steam reforming of propane and oxidation of the generated hydrogen and carbon monoxide in a SOFC, see Figure 8.6. Propane has the advantage that it has a high intrinsic energy density and can be stored as a liquid under moderate pressure. Moreover, the vapor pressure of propane is sufficient to overcome pressure losses and no compressor or pump is needed for the fuel. Atmospheric air is used for the oxidation reactions with an energetic penalty as described above. To overcome heat losses the fuel cell effluent is oxidized and, since the SOFC operates at a high temperature, the air excess from the fuel cell is used for the burner. The residence time in the reactor is assumed to be 2 ms, the operating temperature 1000 K and the outlet temperature 1000 K. A stoichiometric mixture of propane and water is used. The power generated in the SOFC is based on equivalent hydrogen production [2, 35], that is, it is assumed that CO and C_3H_8 undergo internal reforming and that only hydrogen is electrochemically oxidized.

The fourth alternative considered is partial oxidation of propane and oxidation of the generated hydrogen and carbon monoxide in a SOFC, see Figure 8.7. Atmospheric air is used for the oxidation reactions with an energetic penalty as described above. The exothermicity of the partial oxidation reaction suffices to overcome the heat losses. The residence time in the reactor is assumed to be 1 ms, the operating temperature 1000 K and the outlet temperature 1000 K. The power generated in the SOFC is again based on equivalent hydrogen production [2, 35].

The results of the comparison are summarized in Table 8.1. There are several interesting points to note. The partial oxidation of propane leads to the highest energy densities, despite the lowest energy conversion efficiency, while the direct oxidation of methane leads to the highest energy conversion efficiency but

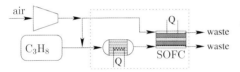

Figure 8.7 Flow-sheet for partial oxidation of propane.

Table 8.1 Results for the comparison of processes.

Performance metric	NH$_3$	CH$_4$	C$_3$H$_8$ ref.	C$_3$H$_8$ POX
Energy efficiency	27%	30%	23%	16%
Volumetric fuel energy density (Wh l^{-1})	1570	170	1740	1920
Gravimetric fuel energy density (Wh kg^{-1})	2580	1610	2540	3900
Volumetric system energy density (Wh l^{-1})	1510	130	1650	1800
Gravimetric system energy density (Wh kg^{-1})	2370	320	2320	3340

the lowest system energy densities. Both the system and fuel volumetric energy densities of methane direct oxidation are very low, due to the large volume required for the storage of the gases (methane and oxygen). Moreover, the gravimetric fuel energy density of methane is low, due to the mass of oxygen which is accounted for [2]. The large difference in gravimetric energy density of methane direct oxidation is due to the mass of the gas cartridges. Despite lower energy conversion efficiency, the propane-based processes lead to higher energy densities than the ammonia-based process, because of the intrinsic difference in energy density between the two fuels. The comparison of the two idealized possibilities for propane fuel processing reactions shows that energy conversion efficiency is not a suitable metric for man-portable applications, because it does not account for the water weight and volume; the higher energy efficiency of steam reforming is due to the generation of additional hydrogen in the reactor. With the exception of the process requiring gas storage, the fuel energy density and system energy density give the same qualitative comparison among processes.

8.4.2
Effect of Scale on Process Performance

The scalability of micropower generation devices is particularly interesting since there are two major scales. One is the nominal power output, which is mainly associated with the device size, and the other is the time between refueling (mission duration), which is associated with the fuel cartridge size. In this case study the influence of these two scales on achievable system performance is presented using the volumetric and gravimetric energy densities as metrics. The parameter values in Table 8.2 are used. The performance is measured using the volumetric and gravimetric system energy densities, where the system includes the power generation device and the stored fuel. Figure 8.8 shows the achievable energy density in Wh/(l system) and Wh/(kg system).

For low power outputs the heat losses dominate over the exothermicity of the fuel processing and burning of the fuel cell effluents as well as part of the fuel is needed (Design I). Since the heat generation scales linearly with power output while heat losses scale sublinearly (with a power of 2/3) the achievable energy density increases significantly with the power output. At a power output of about 0.6 W a kink is observed, because for higher power output the heat generation

Table 8.2 Process parameters for the scaling effect in Figure 8.8.

Property	Value
Ambient temperature	$T_{amb} = 298\,K$
Power output	$PW = 1\,W$
Reactor temperature	$T_{op} = 1000\,K$
Reactor outlet temperature	$T_{out} = 500\,K$
Conversion in reactor	$\zeta = 0.9$
SOFC temperature	$T_{op} = 1000\,K$
Residence time in reactor	$\tau = 1\,ms$
Discard temperature from SOFC	$T_{out} = 500\,K$
Conversion in burners	$\zeta = 0.95$
Residence time in burners	$\tau = 1\,ms$
Air excess in burners	$\Phi = 1.2$
Conversion in fuel cell	$\zeta = 0.8$
Overall heat loss coefficient	$U = 3\,W\,m^{-2}\,K^{-1}$
Residence time in fuel cell	$\tau = 20\,ms$
Emissivity (incl. view factor)	$\varepsilon = 0.2$
Efficiency of fuel cell	$\eta_{FC} = 0.7$
Air excess in fuel cell	$\Phi = 1.2$
Compression parameter for the air feed	$K_C = 10\,J\,mol^{-1}\,K^{-1}$
Burner temperature	$T_{op} = 1000\,K$
Discard temperature from burner	$T_{out} = 500\,K$
Water factor in fuel cell	$\Psi = 1$
Propane molfraction in feed	0.5
No air excess in reactor	$\Phi = 1.0$
Volume factor	10
Fuel cartridge thickness	1 mm
Fuel cartridge density	$1.5\,kg\,l^{-1}$
Start-up time	$\tau_{startup} = 60\,s$
Device density	$\rho = 1\,kg\,l^{-1}$
Auxiliary battery	$200\,W\,h\,kg^{-1}$

from burning the fuel cell effluents is sufficient (Design II). Above approximately 1.6 W the process is sufficiently exothermic, so that the fuel cell effluents need not be oxidized (Design III). The system energy density increases with mission duration and approaches the energy density with respect to the fuel volume/mass because the device size becomes negligible. This case study demonstrates that the influence of scale on process performance is significant. Since, in general, different processes scale differently, the optimal design is also likely to be influenced by the scale.

8.4.3
Effect of Fuel Combinations and Layout

In this case study the effect of using a secondary fuel for heat generation is investigated together with how different layouts can yield significantly different system

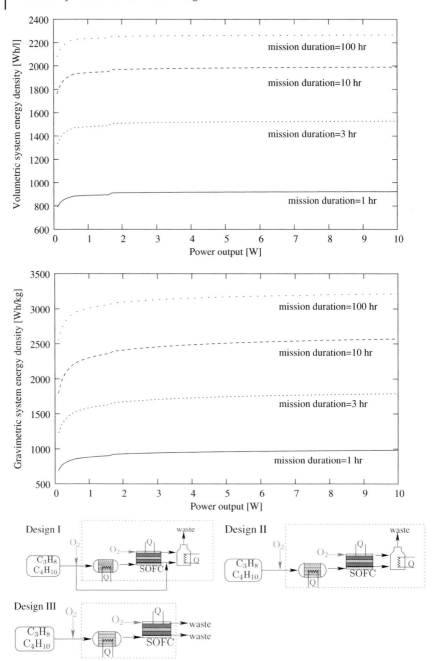

Figure 8.8 Volumetric and gravimetric system energy density
of hydrocarbon partial oxidation in combination with a SOFC
as a function of mission duration and power output.

performances. A process that has been proposed, for example [31], is ammonia decomposition to nitrogen and hydrogen and subsequent oxidation of the hydrogen in a PEM fuel cell. A major drawback of this process is that ammonia is corrosive and extremely toxic. From a technological point of view this process has the benefit that the ammonia does not contain carbon, and thus poisoning of the PEM can be avoided without the need for a separation following the fuel processing. However, it was recently observed that ammonia residuals can lead to a significant performance decrease of PEM [32, 33]. The ammonia-based process has many drawbacks, including high operating temperatures for the fuel processing reactor and an endothermic fuel processing reaction, so that burning the fuel cell effluents may not provide sufficient heat [9]. Performance improvements can be achieved by the use of a secondary high-energy fuel, for example hydrocarbons, for heat generation. Two extreme cases of layout are considered: (i) the two burners are separate and remote heat exchange is used or (ii) the two streams to be burned are combined in a burner which is in thermal contact with the reactor. In [36] it was demonstrated that the conversion for the ammonia cracking reaction is essentially complete for residence times of the order of ms and a fuel processing temperature of 650 °C. Here, complete conversion of ammonia in the reactor is assumed and the residence time in the reactor is varied in the range 0–100 ms. Table 8.3 summarizes the parameters used and Figure 8.9 shows the results in terms of the fuel energy density.

Table 8.3 Process parameters for the ammonia cracking case study, Figure 8.9.

Ambient temperature	$T_{amb} = 298\,K$
Power output	$PW = 1\,W$
Reactor temperature	$T_{op} = 923\,K$
Discard temperature from reactor	$T_{out} = 623\,K$
Conversion in burners	$\zeta = 0.95$
PEM temperature	$T_{op} = 350\,K$
Residence time in burners	$\tau = 1\,ms$
Discard temperature from PEM	$T_{out} = 350\,K$
Air excess in burners	$\Phi = 1.2$
Conversion in fuel cell	$\zeta = 0.8$
Overall heat loss coefficient	$U = 3\,W\,m^{-2}\,K^{-1}$
Residence time in fuel cell	$\tau = 20\,ms$
Emissivity (incl. view factor)	$\varepsilon = 0.2$
Efficiency of fuel cell	$\eta_{FC} = 0.7$
Air excess in fuel cell	$\Phi = 1.2$
Compression parameter for the air feed	$K_C = 10\,J\,mol^{-1}\,K^{-1}$
Burner temperature	$T_{op} = 1000\,K$
Discard temperature from burner	$T_{out} = 500\,K$
Temperature loss factor	$\chi_{temp} = 0.6$
Heat loss factor	$\chi_{heat} = 0.6$

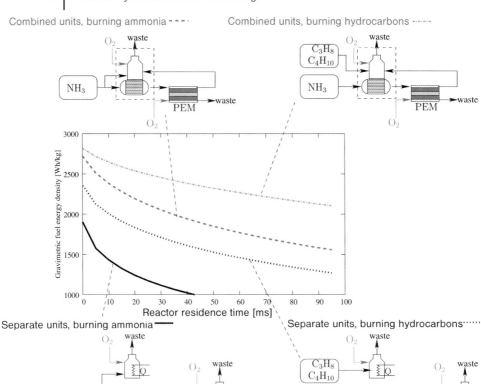

Figure 8.9 Effect of fuel combinations and layout options on gravimetric fuel energy density of an ammonia-cracking based process.

Even for low residence times combining units into a stack has a significant impact and the thermal integration seems necessary. This case study illustrates that flow-sheet design and thermal management, including combination of heat sources and heat sinks, need to be considered simultaneously. The use of ammonia oxidation for heat generation in separate units becomes essentially impossible for high residence times because of the resulting increase in heat losses. The choice of a single fuel or a fuel combination is not obvious; from a perspective of maximizing the energy density, fuel combination is very advantageous, but it bears the logistic difficulties of carrying two fuels. This trade-off implies that for different applications a different design will be used.

8.5
Nomenclature

Abbreviation	Property	Unit
C_{Pi}	Molar heat capacity of species i	$J\,mol^{-1}\,K^{-1}$
e_{grav}	Gravimetric energy density	$W\,h\,kg^{-1}$
e_{vol}	Volumetric energy density	$W\,h\,l^{-1}$
η	Conversion efficiency	$W\,W^{-1}$
F	Faraday constant	$C\,mol^{-1}$
H_i^g	Molar enthalpy of species i, ideal gas	$J\,mol^{-1}$
H_i^f	Molar enthalpy of formation of species i	$J\,mol^{-1}$
H_i^{heat}	Heating value of species i	$J\,mol^{-1}$
ΔH_i^{vap}	Vaporization enthalpy of species i	$J\,mol^{-1}$
$\Delta_r H$	Enthalpy of reaction r	$J\,mol^{-1}$
G_i^o	Molar gas phase Gibbs free energy of pure species i at reference pressure	$J\,mol^{-1}$
I	Current	A
I	Set of species	–
M	Mass	kg
MV_i	Molar volume of species i	$l\,mol^{-1}$
MW_i	Molecular weight of species i	$kg\,mol^{-1}$
$N_{i,j}$	Molar flowrate of species i of stream j	$mol\,s^{-1}$
P	Pressure	bar
P_{ref}	Reference pressure	bar
PW	Power	W
R	Gas constant	$J\,mol^{-1}\,K^{-1}$
T	Temperature	K
T_{amb}	Ambient temperature	298 K
T_{ref}	Reference temperature	298 K
U	Voltage	V
U_{loss}	Overall heat transfer coefficient	$W\,m^{-2}\,K^{-1}$
V	Volume	m^3
z	Number of electrons exchanged	–
ε	Product of emissivity and view factor	$W\,m^{-2}/W\,m^{-2}$
η_{SOFC}	SOFC efficiency	W/W
ζ_r	Conversion of reaction r	$mol\,mol^{-1}$
ξ_r	Extent of reaction r	$mol\,s^{-1}$
ρ_i	Density of species i	$kg\,m^{-3}$
σ_A	Maximal allowable tensile stress	Pa
τ	Residence time	s
$\tau_{mission}$	Mission duration (time between refueling)	h
Φ	Air ratio	$mol\,s^{-1}/mol\,s^{-1}$

Acknowledgments

This work was supported by the DoD Multidisciplinary University Research Initiative (MURI) program administered by the Army Research Office under Grant DAAD19-01-1-0566.

We would like to acknowledge Klavs F. Jensen and the other members of the MIT μChemPower team for fruitful discussions and their input in the formulation of the process alternatives.

References

1 Biegler, L.T., Grossmann, I.E. and Westerberg, A.W. (1997) *Systematic Methods of Chemical Process Design*, Prentice Hall, New Jersey.

2 Mitsos, A., Hencke, M.M. and Barton, P.I. (2005) Product engineering for man-portable power generation based on fuel cells. *AIChE Journal*, **51** (8), 2199–219.

3 Chachuat, B., Mitsos, A. and Barton, P.I. (2005) Optimal design and steady-state operation of micro power generation employing fuel cells. *Chemical Engineering Science*, **60** (16), 4535–56.

4 Dyer, C.K. (1999) Replacing the battery in portable electronics. *Scientific American*, **281** (1), 88–93.

5 One laptop per child http://laptop.org/, May 29, 2008.

6 Mitsos, A., Chachuat, B. and Barton, P.I. (2007) What is the design objective for portable power generation: efficiency or energy density? *Journal of Power Sources*, **164** (2), 678–87.

7 Linden, D. (2001) *Handbook of Batteries*, McGraw-Hill, New York.

8 Dyer, C.K. (2002) Fuel cells for portable applications. *Journal of Power Sources*, **106** (1–2), 31–4.

9 Mitsos, A., Palou-Rivera, I. and Barton, P.I. (2004) Alternatives for micropower generation processes. *Industrial and Engineering Chemistry Research*, **43** (1), 74–84.

10 Powell, M.R., Chellappa, A.S. and Vencill, T.R. (2004) Compact fuel cell power supplies with safe fuel storage. http://handle.dtic.mil/100.2/ADA433359, May 29, 2008 accessed.

11 Underwriters Laboratories Inc. (UL) (1999) http://www.dianyuan.com/bbs/u/28/1113526063.pdf, May 29, 2008 accessed.

12 Meyers, J.P. and Maynard, H.L. (2002) Design considerations for miniaturized PEM fuel cells. *Journal of Power Sources*, **109** (1), 76–88.

13 Saha, N. and Rinard, I.H. (2000) Miniplant design methodology: A case study manufacture of hydrogen cyanide, in *IMRET 4*, Atlanta, 327–33.

14 Franz, A.J., Jensen, K.F. and Schmidt, M.A. (1999) Palladium membrane microreactors, in *IMRET 3*, Frankfurt, Germany, 267–76.

15 Shu, J., Grandjean, B.P.A., Vanneste, A. and Kallaguine, S. (1991) Catalytic palladium-based membrane reactors – a review. *Canadian Journal of Chemical Engineering*, **69** (5), 1036–60.

16 Bullen, R.A., Arnot, T.C., Lakeman, J.B. and Walsh, F.C. (2006) Biofuel cells and their development. *Biosensors & Bioelectronics*, **21** (11), 2015–45.

17 Dyer, C.K. (1990) A novel thin-film electrochemical device for energy conversion. *Nature*, **343** (6258), 547–8.

18 Coors, W.G. (2003) Protonic ceramic fuel cells for high-efficiency operation with methane. *Journal of Power Sources*, **118** (1–2), 150–6.

19 Rinard, I. (1998) Miniplant design methodology, in *IMRET 2*, New Orleans, 299–312.

20 Tester, J.W. and Modell, M. (1997) *Thermodynamics and Its Applications*. Prentice Hall International Series in the Physical and Chemical Engineering Sciences, 3rd edn, Prentice Hall, New Jersey.

21 Delsman, E.R., Uj, C.U., de Croon, M.H.J.M., Schoute, J.C. and Ptasinski, K.J. (2006) Exergy analysis of an integrated fuel processor and fuel cell (FP-FC) system. *Energy*, **31** (15), 3300–9.

22 Delsman, E.R. (2005) Microstructured Reactors for a Portable Hydrogen Production Unit. PhD thesis, Technische Universiteit Eindhoven.

23 Papalexandri, K.P. and Pistikopoulos, E.N. (1996) Generalized modular representation framework for process synthesis. *AIChE Journal*, **42** (4), 1010–32.

24 Horn, J.F.M. (1964) Attainable and nonattainable regions in chemical reaction technique, in *Chemical Reaction Engineering (Proceedings of the Third European Symposium)*, Pergamon Press, London, pp. 1–10.

25 Glasser, D., Hildebrandt, D. and Crowe, C. (1987) A geometric approach to steady flow reactors: the attainable region and optimization in concentration space. *Industrial & Engineering Chemistry Research*, **26** (9), 1803–10.

26 Feinberg, M. and Hildebrandt, D. (1997) Optimal reactor design from a geometric viewpoint: 1. Universal properties of the attainable region. *Chemical Engineering Science*, **52** (10), 1637–65.

27 Bagajewicz, M.J., Pham, R. and Manousiouthakis, V. (1998) On the state space approach to mass/heat exchanger network design. *Chemical Engineering Science*, **53** (14), 2595–621.

28 Mitsos, A. (2006) Man-portable power generation devices: Product design and supporting algorithms, http://yoric.mit.edu/download/Reports/MitsosThesis.pdf. PhD thesis, Massachusetts Institute of Technology.

29 Mitsos, A., Chachuat, B. and Barton, P.I. (2007) Methodology for the design of man-portable power generation devices. *Industrial and Engineering Chemistry Research*, **46** (22), 7164–76.

30 Arana, L.R., Baertsch, C.D., Schmidt, R.C., Schmidt, M.A. and Jensen, K.F. (2003) Combustion–assisted hydrogen production in a high-temperature chemical reactor/heat exchanger for portable fuel cell applications, in *12th International Conference on Solid-State Sensors, Actuators, and Microsystems (Transducers 03)*, Boston, MA.

31 Metkemeijer, R. and Achard, P. (1994) Ammonia as a feedstock for a hydrogen fuel cell; reformer and fuel cell behaviour. *Journal of Power Sources*, **49** (1–3), 271–82.

32 Soto, H.J., Lee, W.K., Van Zee, J.W. and Murthy, M. (2003) Effect of transient ammonia concentrations on PEMFC performance. *Electrochemical and Solid State Letters*, **6** (7), A133–5.

33 Halseid, R., Vie, P.J.S. and Tunold, R. (2006) Effect of ammonia on the performance of polymer electrolyte membrane fuel cells. *Journal of Power Sources*, **154** (2), 343–50.

34 Sifer, N. and Gardner, K. (2004) An analysis of hydrogen production from ammonia hydride hydrogen generators for use in military fuel cell environments. *Journal of Power Sources*, **132** (1–2), 135–8.

35 Science Applications International Corporation (2000) *Fuel Cell Handbook*, 5th edn, EG&G Services, Parsons, Inc.

36 Deshmukh, S.R., Mhadeshwar, A.B. and Vlachos, D.G. (2004) Microreactor modeling for hydrogen production from ammonia decomposition on ruthenium. *Industrial and Engineering Chemistry Research*, **43** (12), 2986–99.

9
Structural Considerations

Brian L. Wardle and S. Mark Spearing

9.1
Introduction

Structural design for micropower devices requires several additional considerations beyond typical microelectromechanical systems (MEMS) structures, but the
basic design philosophy is unchanged from typical deterministic design approaches
used for both MEMS and macroscale structures. Such approaches define a primary
objective for the device, and the structural design proceeds to fulfill (or sometimes
exceed) the objective, often within a long list of constraints. An example primary
objective would be for an accelerometer to provide 1-axis accelerations over a frequency band for the lifetime of an application, perhaps 20 years for an automotive
part. Frequently such design proceeds by ad hoc selection of the device geometry,
followed by sizing the structure to avoid failure using a maximum stress type
failure criterion. A more rigorous approach to preliminary design from a materials
standpoint involves "materials selection", as formalized by Ashby [1]. In this
approach, an objective function is written, usually employing structural mechanics
models, which reflects device functional constraints (e.g. a tensile member with
specified stiffness), a grouping of geometric terms, and a material property grouping. This is the approach adopted here as the groupings of material properties are
instructive for materials selection choices in MEMS power devices, and also
because the formalism allows scaling effects to be revealed straightforwardly. What
is ignored in adopting this approach is the mechanistic understanding of materials
science focused on process structure relationships (see e.g. [2–4]), as the focus is
simply on processing effects on film/device properties.

With the adoption of the materials-selection approach, we must discuss what is
different in MEMS devices [5, 6] from macroscopic structural design, and then
what is different in micropower devices beyond MEMS considerations. Scale
effects [5, 7], strong coupling of processing to materials properties (especially
residual stresses), limitations in both achievable shapes and available materials to
design with, and difficulties in determining quantitatively even basic properties of
(especially new) materials [8], and a relatively high tolerance for failure (compare
airplane damage tolerant design, where all structures must not only survive, but

Microfabricated Power Generation Devices. Edited by Alexander Mitsos and Paul I. Barton
Copyright © 2009 WILEY-VCH Verlag GmbH & Co. KGaA, Weinheim
ISBN: 978-3-527-32081-3

survive when damaged, as opposed to device yield for MEMS devices) set MEMS structural design apart from macroscopic design. Beyond these considerations for MEMS structures, micropower devices present myriad additional challenges, many of them specific to the type of micropower device, for example build-up of carbon on high-temperature solid oxide MEMS fuel cells is very specific to that application. Due to the multiple application-specific challenges, we adopt the approach of focusing on thermomechanical design as many micropower devices involve significant thermal excursions due to processing and/or operation, and therefore design to accommodate significant intrinsic and extrinsic stresses needs to be addressed. Thus, we will focus on thermomechanical design and include examples based on direct experience with MEMS solid oxide fuel cells, microcombustors, microturbines, microrockets, and energy harvesters (see Figure 9.1). Additional general micropower device topics that will be addressed include the introduction of new materials into the available MEMS material set to achieve micropower devices. Examples include co-sputtered cermets for ultra-thin fuel cells, and SiC for microturbines. New materials issues include limited deposition

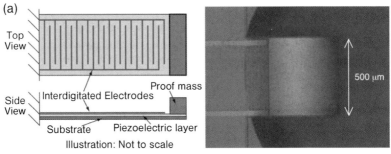

Piezoelectric Vibration Energy Harvester

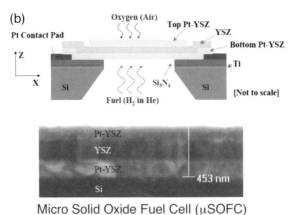

Micro Solid Oxide Fuel Cell (μSOFC)

Figure 9.1 Exemplary micropower devices: (a) thin-film piezoelectric energy harvester, and (b) micro-SOFC. Upper SEM image courtesy of Dr. WooSik Kim (MIT Acro/Astro).

techniques and knowledge, limited or absence of ability to etch, high-temperature depositions and unknown properties of the new material being introduced. These challenges are highlighted through the examples, and also through the concurrent development section at the end of the chapter.

9.2
Structural Design Challenges in Micropower Devices

Stress drivers in MEMS devices are generally characterized as consisting of intrinsic and extrinsic components. While strict definitions are not generally accepted, it is generally accepted that external stress drivers such as mechanical loading and thermally-induced stresses (sometimes treated as *mismatch strains*) are extrinsic drivers [2]. Intrinsic stresses originate from film formation, and include mechanisms such as grain coalescence and growth, forward-peened atoms, impurities, and lattice mismatch in epitaxial-type growth. Classification of the stress is important to develop mechanistic understanding and therefore some control of the stresses through processing, but from a design standpoint the mechanical stress, regardless of source, is what must be captured. In thin film materials, there is a very strong coupling between small process variations and film properties, notably residual stress. Intrinsic stress (residual stress after initial processing) can be very large (GPa-range, for both metals and ceramics) relative to design windows that generally require stress control within ~10 MPa for MEMS, and are strongly influenced by process control such as the position in the wafer boat during CVD deposition. These mechanisms and stresses have been extensively studied for microelectronics and MEMS processing of typical materials such as Si and interconnect metals [3, 4]. At least two additional difficulties arise for micropower devices: (i) high-temperature operation and processing introduce extremely high stresses and stress hysteresis/evolution; (ii) quantitative and mechanistic understanding of intrinsic and extrinsic stresses are not known for new materials (or old materials with new processes), and are strongly process dependent. The latter causes great difficulty in thermomechanical design as basic inputs to any mechanics analysis are not known. Further, stresses and other material properties can evolve with operation and processing. The stress plots in Figure 9.2 help to illustrate both of these points: film stresses vary dramatically based on sputtering gas pressure and location. Note that in both cases stresses vary by more than 1 GPa – these are very high stresses. Processing conditions and sequencing have a strong influence on fabrication flow and for new materials, such as yttria-stabilized zirconia (YSZ) studied in recent fuel cell work, these high stress states are unknown.

Stresses such as those discussed above can lead to dramatic structural failures (see Figure 9.3 examples from micropower device work), including film cracking, device and film buckling, debonding, and other less dramatic device failures such as out-of-plane bending or buckling (e.g. in interdigitated "Comb"-type sensors and actuators which render them structurally viable but useless). Other failure modes include piezoelectric depoling, coking or other surface reactions in micro-

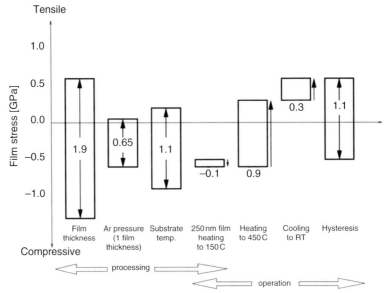

Figure 9.2 Stresses in ceramic YSZ thin film as a function of processing and operation [9–12].

chemical devices, fatigue (uncommon in general for MEMS devices [5], but more likely for high-temperature applications), and electromigration, among others.

The underlying mechanisms that cause residual stresses and hysteresis also lead to process-dependent material properties, in many cases the process structure–property relationships are not straightforward, see for example [13]. Process-dependent properties, such as material stiffness, must be carefully considered as such properties are essential for thermomechanical design. In general, it should be assumed that thin-film properties differ significantly from bulk properties and, therefore, must be characterized. However, this can be extremely difficult as process parameters are numerous and may not be easy to control, can have coupling, and have significant effect on device material properties and therefore device design. An example from our own work on thermomechanical design of MEMS fuel cells with YSZ illustrates these points. Various sputtering conditions were considered for YSZ to achieve the correct phase composition for enhanced electrochemical performance. Along with electrochemical property modification, residual stresses also changed significantly within the processing windows of temperature, pressure, and film thickness studied, giving stresses in the range +100 to −1400 MPa. Note that bulk YSZ has a failure stress of ~700 MPa, illustrating the often-observed thin-film property of increased strength. Further, the modulus of thin-film YSZ was found to be in the range 25–100 GPa (variability due to both processing and test type accounts for the range), whereas the bulk modulus is 200 GPa [11]. Such challenges will be further considered in terms of process sequencing in Section 9.3 on concurrent design. Other considerations

(a)

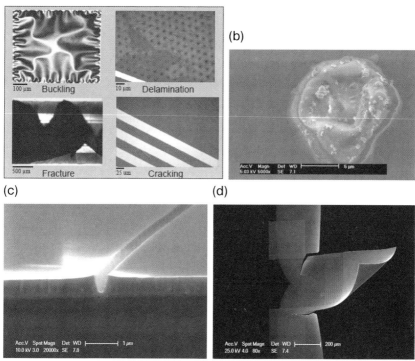

Figure 9.3 Examples of film and device failures. (a) Several stress-driven failures during μSOFC development, (b) contamination failure from a dust particle leading to a shorted fuel cell [11], (c) channel crack in thin-film PZT associated with voids at the substrate interface, and (d) an energy harvester failed due to residual film stresses during development. Lower images courtesy of Dr WooSik Kim (MIT Aero/Astro).

beyond process-dependent properties when working with thin-film materials include patterning and adhesion. These issues tend to be material specific, but an underlying driver is that many of the materials are high-temperature metals or ceramics and are not straightforward to pattern, particularly if the processing involves high temperatures that negates the use of many photoresist processes. An example is again the YSZ electrolyte material which as an inert oxide ceramic is extremely difficult to etch.

9.3
Thermomechanical Design Approach

As discussed in the introduction, micropower device design benefits from following the rigor of "materials selection" [1] using simple mechanical models. Simple models are appropriate due to the high uncertainty and variability associated with material, and also geometric, inputs into the structural models. The design flow may be thought of as follows:

1. Basic structural configuration choice, for example decision on whether it is a beam resonator, a clamped membrane, or a rotating combustion element. This is typically rather straightforward given the device/application.

2. Basic structural mechanics model of the device linking geometry, material properties, and mechanical or thermal drivers of stress.

3. Extraction of material and geometric parameter groupings that are important to maximize or minimize based on the device objective function.

4. Selection of materials that best optimize the material parameter grouping from step 3.

5. Sizing of the structure given expected ranges of material properties, residual stresses, and any drivers of stress (mechanical, thermal, etc.) using basic or refined mechanics models and a simple failure criterion such as maximum stress (or strain) criterion.

6. Fabrication of films and test structures to give better ranges of material properties and residual stresses given processing, and to identify secondary design issues such as adhesion or etching.

7. Iteration between steps 5 and 6 to find a suitable "recipe" for the device in question.

Steps 2–4 are captured in the formalism of materials selection discussed in detail in the book by Ashby [1]. Steps 6 and 7 are fraught with experimental difficulties and take significant investment of time and resources. These last steps are often, but should not be, undertaken without the guidance provided in steps 2–5. Last, the overall recipe in step 7 generally requires some concurrent design, as will be illustrated in Section 9.3 using some examples from our work .

While the full details of the materials selection process are left to the reader to find in the book by Ashby, several of the steps will be illustrated briefly through examples. Consider first a simple device: a beam that we wish to act as a high frequency resonator. Using simple beam mechanics, we can formulate the natural frequency of the beam as a function of geometric and material properties (see Figure 9.4) and then complete steps 1 and 2. To maximize the natural frequency, we quickly see that the material grouping to maximize is modulus over density, E/ρ, that is the specific stiffness. It is intuitive that we would want high stiffness and low mass (density) to maximize frequency. Step 4 for a macroscopic design would involve a very useful tool called a materials selection chart, also due to Ashby. These charts plot various physical parameters against each other for all engineering materials commonly used. An example chart from recent work on carbon nanotubes (CNTs) is shown in Figure 9.5 as an example. CNTs would be seen as having the high specific stiffness, and indeed they have been researched as high-frequency resonators. However, materials selection charts do not exist

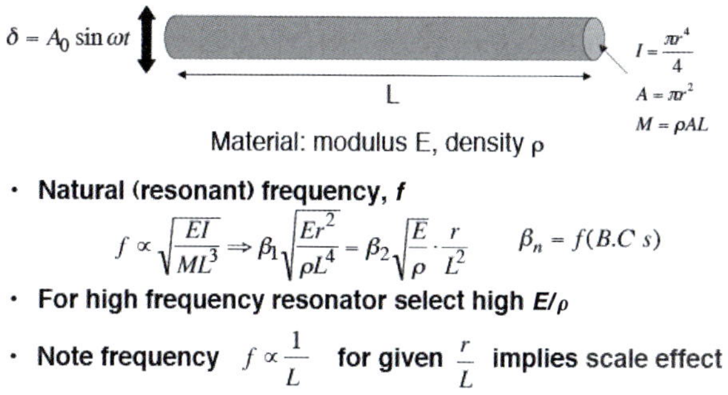

$\delta = A_0 \sin \omega t$

$I = \dfrac{\pi r^4}{4}$

$A = \pi r^2$

$M = \rho A L$

L

Material: modulus E, density ρ

- **Natural (resonant) frequency, *f***

$$f \propto \sqrt{\frac{EI}{ML^3}} \Rightarrow \beta_1 \sqrt{\frac{Er^2}{\rho L^4}} = \beta_2 \sqrt{\frac{E}{\rho}} \cdot \frac{r}{L^2} \qquad \beta_n = f(B.C\ s)$$

- **For high frequency resonator select high *E/ρ***

- **Note frequency** $f \propto \dfrac{1}{L}$ **for given** $\dfrac{r}{L}$ **implies scale effect**

Figure 9.4 Analysis to derive the material and geometric parameters in the design of a simple beam resonator. β is a function of the beam boundary conditions (BCs).

currently for thin-film materials. As mentioned earlier, many physical parameters such as strength, yield stress, and modulus can be very different in thin-film form than in bulk form. Thus, for MEMS devices, one needs to turn to the best-available tabulated data such as that shown in Table 9.1 for some common MEMS materials. Si performs very well on the specific stiffness metric, and better still are SiN, alumina, SiC, and diamond, respectively. Note that these materials are high-performance choices even considering macroscopic advanced materials, high-strength steel and advanced carbon fiber composites have specific stiffness of ~25 and 88 MN-m kg^{-1}, respectively. Finally, although not specific to micropower devices, the common scale effect noted for MEMS and NEMS in the analysis in Figure 9.4 should be noted. Everything else being equal (geometric ratio r/L constant), as the beam resonator is scaled down, the frequency increases. As CNTs have both high

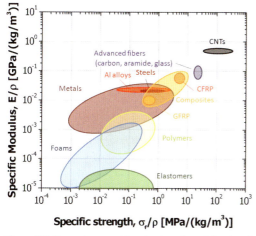

Figure 9.5 Example materials selection chart following the work of Ashby, courtesy of Dr. Roberto Guzman deVilloria (MIT Aero/Astro).

Table 9.1 Materials grouping parameters for common MEMS materials related to the examples given in the text. Note that these values will vary greatly depending on processing specifics.

Material	Modulus, E GPa	Density, ρ kg m^{-3}	Useful Strength[a], σ MPa	E/ρ MN-m/kg	$\sigma_f/E \times 10^{-3}$	σ_f/ρ MN-m kg^{-1}
Silicon	165	2330	4000	72	24	1.7
Silicon oxide	73	2200	1000	36	13	0.45
Silicon nitride	304	3300	1000	92	3	0.30
Nickel	207	8900	500	23	2	0.06
Aluminum	69	2710	300	25	4	0.11
Aluminum oxide	393	3970	2000	99	5	0.50
Silicon carbide	430	3300	2000	130	4	0.303
Diamond	1035	3510	1000	295	1	0.28

a   Strength data is highly variable – intrinsically and due to processing effects, particularly in thin films.

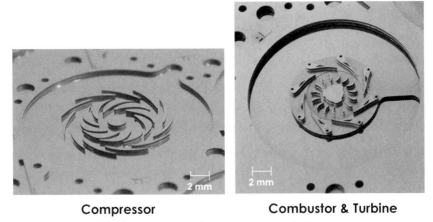

Compressor   **Combustor & Turbine**

Figure 9.6 Microturbine devices, courtesy the MIT Microengine Project.

specific stiffness and are one of the smallest beams known, this gives further insight into why they are investigated as ultra-high frequency resonators.

Building on this simple example, three micropower component examples will be introduced briefly: a MEMS fuel cell designed to resist failure due to buckling, a microturbine rotor designed for power output, and a piezoelectric energy harvester designed so the transduction properties do not degrade due to cyclic loading (fatigue). The fuel cell and energy harvester have been introduced previously (see Figure 9.1) and microturbine elements are shown in Figure 9.6. In the first example of a fuel cell, a freestanding plate (see Figure 9.1) is designed to resist large in-plane stresses as well as buckling [13]. The materials in the three-layer plate (anode, electrolyte, and cathode) are chosen for reasons of electrochemical performance, rather than from a materials selection analysis. The design space can be constructed in the two dimensions of applied thermal change (which induces stresses due to coefficient of thermal expansion (CTE) mismatch with the substrate) and total residual stress due to processing. In this way, a design window emerges giving a range of residual stresses that can be withstood, as well as temperature changes that can be tolerated before buckling [13]. In recent work, a significant expansion of the design window was achieved by considering design into the nonlinear postbuckling regime [11]. Next, consider an unusual rotating MEMS structure: a microturbine rotor (see Figure 9.6). In this case, the materials selection in steps 2–4 to maximize power density yields a critical grouping of materials parameters of failure stress over density (σ_f/ρ). Referring to Table 9.1, Si is noted to have a number of advantages. However, if high-temperature operation is required, the retention of strength at temperature becomes a dominant design consideration. Given the lack of good microfabrication processes for refractory materials, inevitably a compromise is reached. In the case of the Microengine project, this compromise has been to use local reinforcement with CVD SiC, while defining all the critical dimensions and aerodynamic surfaces in Si [14].

Finally, a mechanical vibration energy harvester is considered. These are layered devices, with a critical component being a piezoelectric ceramic film (usually lead-zirconium-titanate, PZT) that converts mechanical strain into electrical charge. The energy harvesters are designed for power optimization [15–17], but a critical issue for device operation is that the piezoelectric properties remain intact throughout the life of the device. Bulk piezoelectrics suffer from 'piezoelectric fatigue' where the material loses the ability to transduce over time. Since the operating principle of the device is to convert vibrational energy into electrical energy, fatigue must be considered. Thus, rather than designing to a stress limit, a piezoelectric limit is considered in fatigue. The same models used to design for power can be used to extract the maximum strain and stress in the piezoelectric layer, and the device is assessed for the level of strain the piezoelectric will see. Fortunately, in the designs considered to date, the levels of strain and stress have been below a typical *bulk* value for piezoelectric fatigue. As with many thin-film properties, the piezoelectric fatigue threshold for PZT thin films is not well characterized and is expected to be a strong function of processing specifics.

9.4
Concurrent Development in Micropower Device Design

Earlier in this chapter we outlined many of the various challenges involved in designing and realizing micropower devices. An overview of the basic design approach is provided and illustrated with micropower examples from our own work in the last section. That discussion left most of the challenges unaddressed. Indeed, many of the challenges to device realization occur in step 6 and the iterations suggested in step 7. Here, two aspects of concurrent design are discussed, based on both authors' experience on a large multi-device microchemical power project involving high-temperature reactions of hydrocarbons linked to a fuel cell to generate electrical power. The two aspects of concurrent design that were important based on our experience from a thermomechancial design perspective are:

1. New information: A continuously evolving understanding of properties, and particularly residual stresses, as fabrication with new materials and integration proceeds.

2. New systems-level information: Systems-level analysis can provide strong guidance as to how well a design may evolve to best achieve system-level objectives.

New information provided in both these areas requires the design, and indeed the overall fabrication flow leading to the device, to evolve continuously. Due to the very strong coupling between processing and properties, especially residual stresses, in micropower device materials, basic inputs to thermomechanical design analyses are continuously evolving and the changes are significant. This is in contrast to typical macroscale thermomechanical design, where material properties are well-known and even regulated, as in metallic alloys for aerospace vehicle design. The key difficulty is that materials properties (and the accompanying processes) have a long lead time to develop relative to other aspects of design. In MEMS devices, and particularly in the fuel cell work, properties are not known for the new materials being developed in thin-film form, making design highly uncertain. As films and structures are fabricated, more becomes known, and the design and fabrication flow will likely change. In the case of the fuel cell, the primary driver for device performance was the electrochemical performance of the films. As the correct recipes (sputtering parameters) were discerned for the thin-film YSZ electrolyte, it was necessary to measure residual stresses concurrently and, where possible, determine modulus and CTE parameters as well.

New information also came into play from systems-level analysis of the overall microchemical power system (see Chapter 11 dedicated to optimal design and operation). An example was the finding that, from a systems perspective, the fuel cell should operate above 1000 °C for maximal energy density [18]. The thermomechanical and materials design had been working towards 700 °C or even cooler designs up until that point. Closing the gap between these two operating regimes was not possible in the timeframe of the project, but it did allow us to push for higher-temperature devices as opposed to cooler ones. Systems-level input into a design is not specific to MEMS or MEMS micropower devices but, coupled with the MEMS-specific concurrent design drivers of evolving material properties, it adds additional complexity. In the fuel cell device development, our group was extremely fortunate to have systems-level input: for most projects this valuable information is simply not available. We believe there is significant improvement in overall design cycle time that can be realized with careful attention to the concurrent design issues touched upon in this section.

The topic of thermomechanical design in micropower devices is by no means closed. Challenges unique to microfabrication of micropower devices delineated early in the chapter and exemplified throughout, remain. Perhaps the most significant challenge in the design of such devices is the lack of well-described and repeatable materials properties that are key inputs to any structural design exercise. Macroscale structural systems are rarely developed concurrently with new materials, or conversely, macroscale systems are engineered with existing and well-quantified materials properties. While there is great capability to predict very accurately (primarily through the finite element method, see e.g. [19]) stress and strain states in even complicated geometries, the results of such analyses depend on key inputs such as modulus and residual stress. Typically, such properties are not well-described nor repeatable for materials in thin-film form. This is even more the case in many micropower efforts where exotic or new materials are explored for functionality in thin-film form, for example YSZ and the cermet Pt-YSZ in the case of micro-fuel cells. Certainly, as the MEMS field advances generally, this situation will improve as better control over processes (and therefore properties) emerges with time. The exciting potential for micropower devices, given the numerous applications discussed throughout this book, will help build the required understanding of processes and properties. As the community improves this base of understanding, structural design will be accelerated and more detailed designs will become more practical at earlier stages in device development.

Acknowledgments

The authors would like to thank the MIT MicroEngine Project and the entire MIT MicrochemPower MURI team, including Profs Klavs Jensen, Martin Schmidt, and Harry Tuller, and the students and Postdocs, especially Andrew Ie, David Quinn, Srikar Vengallatore, Kevin Turner, Namiko Yamamoto, Dr Nathan Wicks, and Dr Joshua Hertz. BLW thanks Dr WooSik Kim for the energy harvesting images and discussion, and Dr Roberto Guzman deVilloria for the materials selection chart development.

References

1 Ashby, M.F. (2005) *Material Selection in Mechanical Design*, Pergamon Press, Oxford.

2 Freund, L.B. and Suresh, S. (2003) *Thin Film Materials: Stress, Defect Formation and Suface Evolution*, Cambridge University Press, Cambridge.

3 Ohring, M. (1992) *The Materials Science of Thin Films*, Academic Press, New York.

4 Thompson, C.V. (2000) Structure evolution during processing of polycrystalline films. *Annual Reviews of Material Science*, **30**, 159–90.

5 Madou, M. (2001) *Fundamentals of Microfabrication*, 2nd edn, CRC Press, Boca Raton, FL.

6 Senturia, S.D. (2001) *Microsystem Design*, Kluwer, Norwell MA, USA.

7 Spearing, S.M. (2000) Materials issues in MEMS. *Acta Materialia*, **48**, 179–96.

8 Srikar, V.T. and Spearing, S.M. (2003) A critical review of microscale mechanical testing methods used in the design of microelectromechanical systems. *Experimental Mechanics*, **43** (3), 228–37.

9 Quinn, D., Wardle, B.L. and Spearing, S.M. (2008) Residual stress and microstructure of As-deposited and annealed sputtered Yttria stabilized Zirconia thin films. *Journal of Materials Research*, **23** (3), 609–18.

10 Quinn, D.J. (2006) Microstructure, residual stress, and mechnical properties of thin film materials for a microfabricated solid oxide fuel cell. Master's thesis, Massachusetts Institute of Technology.

11 Yamamoto, N. (2006) Thermomechanical properties and performance of microfabricated solid oxide fuel cell structures. Master's thesis, Department of Aeronautics and Astronautics, Massachusetts Institute of Technology.

12 Choi, D., Shinavski, R.J., Steffier, W.S. and Spearing, S.M. (2005) Residual stress control in thick LPCVD polycrystalline 3C SiC coatings on Si substrates. *Journal of Applied Physics*, **97** (7), 074904.

13 Srikar, V.T., Turner, K.T., Ie, T.Y.A. and Spearing, M.S. (2004) Structural design considerations for micromachined solid-oxid fuel cells. *Journal of Power Sources*, **125**, 62–9.

14 Moon, H.S., Choi, D. and Spearing, S.M. (2004) Development of Si/SiC hybrid structures for elevated temperature micro-turbomachinery. *Journal of Microelectromechanical Systems*, **13** (4), 676–87.

15 duToit, N.E., Wardle, B.L. and Kim, S.-G. (2005) Design considerations for MEMS-scale piezoelectric vibration energy harvesters. *Integrated Ferroelectrics*, **71**, 121–60.

16 duToit, N.E. and Wardle, B.L. (2006) Performance of microfabricated piezoelectric vibration energy harvesters. *Integrated Ferroelectrics*, **83**, 13–23.

17 duToit, N.E. and Wardle, B.L. (2007) Experimental verification of models for microfabricated piezoelectric vibration energy harvesters. *AIAA Journal*, **45** (5), 1126–37.

18 Chachuat, B., Mitsos, A. and Barton, P.I. (2005) Optimal design and steady-state operation of micro power generation employing fuel cells. *Chemical Engineering Science*, **60** (16), 4535–56.

19 Bathe, K.-J. (1996) *Finite Element Procedures*, Prentice Hall, Englewood Cliffs, NJ.

10
Microreactor Engineering: Processes, Detailed Design and Modeling

Dionisios G. Vlachos

10.1
Introduction

Microchemical systems could be central, for different reasons, to the generation of both new portable and distributed power sectors. In the case of portable electronics, there is an ever-increasing demand for more integrated, lighter, and longer lasting devices. Portable electronics, including laptops, iPods, cellular phones, global positioning devices (GPS), unmanned aerial vehicles (UAVs), and telecommunications, rely currently on batteries, which are too heavy and do not last long. The growth of portable power devices necessitates the exploration of lightweight, efficient, and portable power generation devices. The high energy density of most liquid fuels, in comparison to some of the best available batteries, indicates that even a moderately efficient chemical system, which utilizes fuels, could outperform batteries in terms of duration and weight. In addition, chemical systems are more environmentally benign, are easier to recharge by simple refueling, and do not create logistics issues with (partially) unused batteries that have to be left behind.

Construction of systems of different throughput is a fairly straightforward task based on *scale-out or numbering up*, that is one replicates a microsystem multiple times. This scaling feature makes meeting variable power needs a tractable task. Their compact size and scaling-out capability along with the ability of imparting multifunctionality to improve efficiency and emissions constitute a key motivation for employing microchemical technology to enable future power generation. The inherent complexity of this technology, stemming in part from process intensification and multifunctionality, demands a departure from the traditional approach of semi-empirical modeling and traditional process design, control, and optimization. These systems open up new opportunities for research and technological advances.

Intensive research efforts on microreactors and other microchemical systems were initiated in Germany in the late 1980s. Work on modern microreactors in the USA appeared for the first time in the mid 1990s, starting at DuPont and MIT [1, 2] and later at PNNL and Battell and was catalyzed by a related workshop [3].

Initial efforts were based on Si-technology, building on fabrication advances in microelectromechanical systems (MEMS) [4]. However, silicon is limited to fairly low temperatures and, thus, recent efforts have been devoted to metals and ceramics, see for example [5–7]. These advances have been summarized in recent reviews [4, 8–13]. Most of the earlier work on microreactors, however, did not focus on power generation.

In this chapter, various methods for production of power at the small scale are reviewed and compared. In addition, the challenges in designing microreactors and some solutions are discussed. Finally, the need for detailed and simplified simulations is outlined and various examples of microreactors from the author's research group are presented to illustrate key points.

10.2
Process and Catalyst Selection for Microreactors

Process design and optimization have traditionally focused on large manufacturing facilities in order to reduce overall cost. By-and-large, scale-up rules do not apply at microscales [14, 15], as Table 10.1 illustrates. Examples from the conversion of natural gas to hydrogen (combustors integrated with reformers), discussed next, serve to illustrate the need for development of design rules appropriate for small systems.

Power can be produced at small scales using fuels via various means. The most obvious one is to produce hydrogen (H_2) for PEM fuel cells, as shown in Figure 10.1.

Gas-phase combustion that is employed to supply heat to reformers is an excellent manifestation of processes being drastically different at small scales. Flames result in high wall temperatures, preventing viable, extended operation [16], and possess serious safety concerns for portable devices. Catalytic combustion, while not preferred at large scales due to mass transfer limitations, becomes fast at small scales, as a result of process intensification, and is a compelling technology for microdevices. An illustrative example comparing the two technologies at small scales is shown in Figure 10.2.

Attempts to scale down the steam reformer for automobiles (e.g. Chevrolet S10) have been met with moderate success leading to negative views for on-board reforming [17, 18]. It is clear that processes with long residence times, such as reforming, cannot be designed in the same manner as their large scale counterparts. Microreactor technology with fast heat transfer, faster chemistry and more active catalysts are necessary prerequisites to enable operation at smaller scales. For example, change of catalyst (from Ni to Rh) and process intensification by miniaturization may be sufficient for scaling down the steam reforming process [19]. Alternatively, fast chemistry [20–22], such as the catalytic partial oxidation (CPOX) of liquid fuels

$$Fuel + O_2 \rightarrow xCO + yH_2,$$

Table 10.1 Main characteristics, issues and needs for microreactors.

Characteristic	Issues	Needs
Laminar flows	Mixing relies on molecular diffusion only and is slow	Efficient micromixers of low pressure drop
Small size	Difficult to load enough catalyst and ensure complete conversion	Develop deposition schemes and structures to load enough catalyst; high integration of reactor(s) and separation units for compact systems; fast chemistry/process; very active (and selective) catalyst.
Reactors shake in portable devices	Moveable parts break; bypass from pellet settling can occur	Monolithic structures with no moveable parts
High pressure drop	Pressure drop increases with decreasing pellet size	Open (extruded monolith-like) geometries
Transient operation very common	Most designs rely on steady state operation and control; catalysts, which are stable under steady state conditions, may deactivate during start up and shut down; start-up can be slow.	Get heat in and out of the system quickly; develop appropriate designs and strategies; models for dynamics.
Surface to volume ratio is high	High transport rates; high heat losses; thermal management is challenging.	Highly (thermally) integrated systems and minimization of heat loss; novel thermal management strategies.
Compact devices	Match of systems operating at different temperatures.	New detailed models and new systems' design, optimization, and control principles.

gives high fuel conversion in compact devices, that is it scales down well and needs no heat (with obvious consequences for reduced fuel utilization and emissions). As a result, CPOX may be a better fuel processing route for portable and distributed power generation.

Both steam reforming and partial oxidation result in syngas production. If pure H_2 is needed for power, then syngas must be purified. This requires typically two water-gas shift reactors (a high-temperature system followed by a low-temperature system) and a separation unit, such as pressure swing adsorption or a membrane, or a preferential oxidation reactor (PROX). Integration of multiple units operating at different temperatures is obviously a daunting task at the microscale. An alternative fuel-processing route is to dehydrogenate a fuel. In the case of

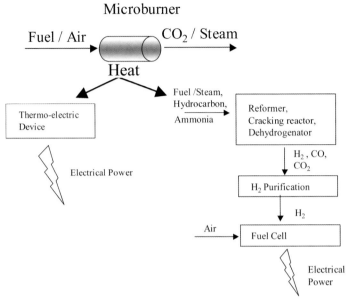

Figure 10.1 Schematic of various paths toward electricity produced from fuels. Microburners supply heat that can be used to drive either thermoelectric elements or an endothermic reactor producing H_2. Depending on the H_2 production route, the H_2 purification step can be fairly simple (e.g. NH_3 cracking) or very involved (e.g. fuel steam reforming or partial oxidation).

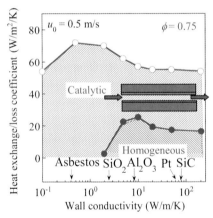

Figure 10.2 Comparison of attainable region of catalytic and homogeneous combustion of premixed C_3H_8/air mixtures under identical conditions (an inlet flow velocity of $0.5\,\mathrm{m\,s^{-1}}$ and an equivalence ratio of 0.75). Results of CFD simulations in terms of the outside heat transfer coefficient for heating an endothermic reaction vs. the wall material conductivity (inset: schematic of a parallel plate geometry with catalyst plates separated by a gap of $600\,\mu m$ in this example). The shaded area under each curve denotes the regime within which operation is feasible. The points denote end operation points above which operation is not sustainable. The maximum heat transfer coefficient in homogeneous combustion is in the natural convection region, that is homogeneous combustion *cannot* be coupled with endothermic reactions. Catalytic microcombustion is much more stable and suitable for microchemical devices.

hydrocarbons, such as ethane, this could lead to coke formation and shut-down of the system. Ammonia decomposition (for example on a Ru catalyst)

$$2NH_3 \rightarrow N_2 + 3H_2, \quad \Delta H_r^\circ = 91.8 \text{ kJ/mol}$$

is an excellent alternative route that produces fairly pure H_2 (with some residual ammonia that needs to be scrubbed prior to feeding the stream to a PEM fuel cell) without poisoning (coking) of the fuel-processing catalyst [23–25]. At high temperatures, reaction goes to completion at millisecond contact times [24, 25]. In addition, it is fairly easy to store ammonia for portable devices, something that is very difficult with small hydrocarbons.

An alternative route to electricity is the integration of a burner with a thermoelectric device [26], as shown in Figure 10.1. This is a much simpler system (and thus more robust at small scales), than the fuel processor/PEM fuel cell route, consisting of two main units: a microburner and a thermoelectric. The latter, however, has a low efficiency. However, when accounting for the efficiencies of both a PEM fuel cell and a fuel processing system (which are moderate to low due to heat losses), a microburner/thermoelectric system may be competitive.

In summary, different processes, catalysts, and chemistries may be needed to render microtechnology viable for power generation. In addition, new reactor configurations are needed to overcome challenges and take advantage of process intensification of microsystems, as outlined below.

10.3
Microreactor Configurations for Portable Power

10.3.1
Need for Novel Processes and Reactor Configurations

Table 10.1 summarizes some of the main differences of microreactors from their large-scale counterparts and highlights challenges and needs in their design. Overall, miniaturization of future fuel processing units demands rethinking of catalysts and chemistries, as discussed above. In order to make microreactors for power viable, one also needs new reactor configurations, for the reasons outlined next.

10.3.1.1 Mixing
In large-scale reactors, the flow is typically turbulent and mixing is fast. In microdevices, on the other hand, the flow is laminar (the Reynolds number is often in the range of 1–100). As a result, mixing is driven by molecular diffusion and is typically slow. Several concepts have recently been explored in the literature including active and static micromixers that rely on laminar shear flow, distributive mixing, hydrodynamic focusing, or vortices (see [27] and references therein). Micromixers (with the exception of interdigital ones) are designed for low through-

put. An array of microposts or pillars of certain size, alignment and spacing can spread the flow to improve contact with the catalyst and enhance mixing due to vortex shading [27]. These static micromixers exhibit a low-pressure drop and, when combined with multilamination, can provide fast mixing at small scales. Figure 10.3 summarizes some of these results.

10.3.1.2 Catalyst Area

For most catalytic reactors, high surface area is needed. It is challenging to put enough catalyst in a small volume to achieve full fuel utilization (conversion nearly 1), especially at short residence times (fast flows). The traditional approach for creating high surface area catalytic reactors encompasses particulates (ceramic pellets of ~1 mm) impregnated with catalyst nanoparticles in, for example, fixed and fluidized bed reactors. For the microreactor analog, one needs to scale down pellets to μm size, resulting in a huge pressure drop. In addition, for portable devices, bypassing, due to settling of particles, can occur. These challenges point toward monolithic types of structures, similar to the widely used automotive catalytic converter. Metal posts (pillars), similar to the ceramic ones shown in Figure 10.3, could be used, by anodization, to create a mesoporous structure for high catalyst loading [23]. Alternatively, one could use a low pressure drop, parallel plate geometry with mesoporous catalyst wafers [29] inserted for high catalyst loading (Figure 10.4). These inserts provide a thousand-fold increase in surface area in comparison to the geometric surface area. A static micromixer near the entrance can be employed to overcome mixing and contact issues, as shown in Figure 10.4. Mesoporous structures strike a good balance between high surface area and low internal mass transfer.

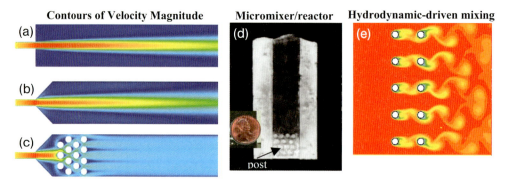

Figure 10.3 (a–c) 2D velocity contours from 3D CFD simulations indicating that for fast (millisecond contact time) flows, straight or injector type of connections to microchannels lead to jetting and bypass of the flow from the majority of the catalyst. (d) Picture of an actual experimental design of a micromixer (bottom) followed by a microreactor section (top part) [6]. (e) 2D CFD simulations indicating that an appropriate design of posts can lead not only to good contact with the catalyst layer, as in (c) and (d), but also to a flow-induced instability resulting in enhanced mixing [28].

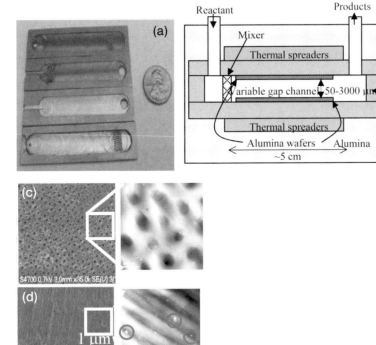

Figure 10.4 (a) Photograph of and (b) schematic side view (not to scale for ease of visualization) of a composite wall, tunable microreactor; (c) scanning electron micrograph (SEM) of a porous anodized alumina wafer ~75 μm thick (top view). The pore diameter is ~50 nm; (d) SEM of pores running across the wafer. Catalyst is deposited within the wafer via dip-coating. Pt nanoparticles are indicated at the right-most TEM images (circles) at the pore mouth but mainly within the pores, and are confirmed by elemental analysis. Micromixers [28] are also embedded within the microreactors.

10.3.1.3 Transients

Many microchemical systems may also be subjected to transient operation. This is admittedly a more challenging task: one needs to get heat in and out sufficiently fast to start-up and shut-down the system, minimize emissions during start-up, and ensure that the catalyst does not deactivate (e.g. due to sintering arising from temperature overshoots in the transients). While significant work has been done at the macroscale on ignition, little is known for the start-up and shut-down of microscale devices [30–32].

10.3.1.4 Safety

Some commercial catalysts deactivate during transients and/or are pyrophoric. An example is Cu-based catalysts in the water-gas shift (WGS) reaction (used to convert syngas to H_2)

$$CO + H_2O \rightarrow CO_2 + H_2, \quad \Delta H_r^\circ = -41 \text{ kJ/mol}$$

that are pyrophoric and deactivate in the presence of air and condensed water. Hence, there has been an on-going interest in noble metal catalysts (e.g. Pt, Rh) that are non-pyrophoric and are highly active [28, 33–35].

10.3.1.5 Materials Selection

With the chemistry taking place on the walls of catalytic microreactors or nearby for gaseous microreactors, there is an intimate coupling between the fluid and the wall. Specifically, due to the large aspect ratio of microreactors, radiation is typically not as important [36], and conduction along the walls is the dominant heat transfer mechanism [37, 38]. As a result, selection of wall materials is very important. Creation of "tunable" microreactors [29], whereby one can attach various thermal spreaders to a thin microreactor framework, allows experimental testing of various effective conductivities for optimal performance.

10.3.2
Process Intensification via Multifunctionality

Industrial processes operate nearly adiabatically, due to their low surface area to volume ratio, and are often separate from each other. The need for compact devices, fast and efficient heat management, and fast transients requires a different strategy [14, 15, 39, 40], an all-in-one systems approach. This naturally leads to multifunctional devices, where one imparts multiple functionalities within a device (e.g. heat exchange, microseparation, multiple reactors). This multifunctionality enables further process intensification by improving heat integration and minimizing heat losses and emissions. However, multifunctionality implies that system rather than individual components' optimization is needed. Very few studies have been devoted to multifunctional microchemical devices and, specifically, to heat management. Next we discuss relevant multifunctional devices that emphasize one of the most critical aspects of miniaturized microsystems, namely energy integration.

10.3.2.1 Integration of Two Reactions via Heat Exchange

Coupling of endothermic and exothermic reactions can be achieved in a number of ways. The concept of carrying reactions on opposite sides of a wall (often referred to as a catalytic wall reactor; Figure 10.5) coated with catalyst on each side has been explored for several years: it allows fast heat transfer via conduction through the walls and eliminates resistance through the boundary layer(s). Tradi-

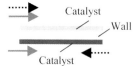

Figure 10.5 Schematic of co-current (same direction arrows) and counter-current (opposite direction arrows) flow configurations for the catalyst wall reactor. Heat liberated from an exothermic reaction conducts rapidly to the endothermic reaction.

tionally, the counter-current heat exchange mode has better performance. Recent work [41–44] on mesoscale devices (gap size > 1 mm) has demonstrated that the co-current flow configuration lowers catalyst temperature due to the overlap of reaction zones.

While the foundations of this multifunctional mode at large scales are now fairly well understood [43], less is known about the extendability of this concept to microsystems. Experimentally, this concept has been shown to lead to substantial reduction of processing time of steam reforming, especially when coupled with introduction of very active, noble metal catalysts [19]. Computational fluid dynamics (CFD) work on microchemical systems (coupled homogeneous combustion with NH_3 cracking on Ru catalyst for H_2 production) underscored differences from larger scales along with the profound role of wall conductivity in the preferred mode of flow configuration [45, 46]. While counter-current operation may give higher efficiency under certain conditions, co-current operation gives lower temperatures, minimizes hot spots, and allows use of most materials. It should thus be preferred, even at the microscale.

10.3.2.2 Recuperative and Regenerative Heat Integration

In order to improve heat integration, "excess enthalpy" geometries have been used in which the heat released from an exothermic reaction is recycled back [47]. Various designs have been considered [48] for spatial thermal coupling. At small scales, it is important to do this compactly to minimize heat loss between components. Two dominant strategies of "excess enthalpy" are reverse flow and heat recirculation. The heat recirculation geometry (Figure 10.6b) utilizes the exiting gas enthalpy by sending the hot products over the combusting reactants counter-currently [49]. This concept has recently been applied, for example, to small-scale burners [50–53] and "swiss-roll" microburners, where the exiting stream is recirculated in a spiral [54–56]. Limited studies have exploited this concept for fuel processing, for example [19, 57].

In reverse flow (Figure 10.6a), the inlet and outlet streams are periodically switched, thus trapping heat within the reactor [58]. This is accomplished in fixed bed reactors by storing energy in solid pellets. Furthermore, it has been limited

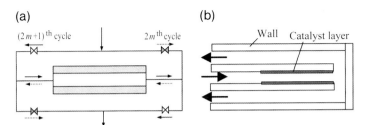

Figure 10.6 (a) Schematic illustrating the working principle of reverse flow operation. The flow direction is reversed by alternately switching on and off the shaded and unshaded valves. (b) Schematic of heat recirculation geometry. The arrows indicate the direction of flow.

to low calorific value fuels (elimination of organics). Only recently, the concept of reverse flow has been shown to substantially improve syngas yield in monoliths with some catalysts (e.g. Pt, Ni) [59, 60]. "Temporal" coupling of two reactions, where the exothermic and endothermic reactions are alternately carried out in the same reaction chamber, is a related approach, see for example [61, 62]. The reverse flow concept works well in gas-phase microreactors wherein the walls store energy [63] leading to substantial improvement on the blowout stability limit but has minimal effect on the extinction limit. Similar improvements are expected in catalytic microchemical systems.

10.4
Multiscale Modeling, Design, and Optimization

CFD, with empirical kinetics and simplifying assumptions (e.g. isothermal walls), is often employed, for example [19] to provide insights for design and fabrication of microreactors. However, the underlying assumptions are often questionable. For example, reaction rates for steam reforming (SR) on Ni catalyst published by various groups differ by orders of magnitude [64]. As a result, the correctness of results (even qualitative trends) is not always clear. Even if the kinetics was fitted, this approach is serious plagued by its lack of reliability for model-based optimization and control.

An alternative strategy is to construct first-principles models, starting from quantum mechanical density functional theory (DFT) and ending up with CFD via multiscale modeling that enables coupling of models across scales [65]. While fundamentally rigorous, this is obviously a daunting and probably not fully justified approach. Clearly, not every detail matters in the design of microchemical systems. Even if multiscale models were somehow developed, their use for optimization of internal microstructures and multifunctional devices is (and will remain to be in the foreseeable future) beyond our computational capabilities.

10.4.1
Hierarchical Multiscale Modeling Framework

The dichotomy of the need for detailed, accurate, complex multiscale models (to ensure good predictions) and model simplicity (to ensure systems tasks, such as optimization) can be reconciled by a new framework [65–68], which is termed hierarchical multiscale modeling. This framework starts with a simple model at each scale (e.g. semi-empirical parameter estimation methods for kinetics, such as the bond-order conservation, transport correlations, etc.) and performs simulations to identify, via sensitivity analysis, which phenomena, scales, and models are crucial ("needed" parameters) [69]. It then refines these key parameters and models, using higher-level tools (e.g. DFT, Monte Carlo (MC), molecular dynamics (MD), CFD). The procedure is iterative and delivers models whose key parameters and scales are resolved with high accuracy whereas others remain approximate.

Upon development, model compression (reduction) is performed over the entire parameter range of interest in order to enable systems tasks, such as optimization and control, to be performed. Key in this approach is the coupling of phenomena and models across scales that enable one to go up and down the multiscale "ladder", that is to predict macroscopic behavior while retaining first principles information [24, 45, 46, 70], to manipulate macroscopic variables to control phenomena at small scales [68, 71], and to design experiments and possibly catalysts based on multiscale models [69].

Reduction of a detailed chemistry model (microkinetic model) is essential to cut down on the number of gaseous and surface species and reduce the number of equations to be solved. While this has been done for several decades, model development was based on a priori assumptions and subsequent fitting (a reductionistic modeling approach). In the new modeling framework, model fitting is minimized; instead parameters are estimated from semi-empirical methods and the key ones are improved using first principles techniques. In addition, no a priori assumptions are made. Instead, the complex model is reduced based on what is important, that is with "right" choices.

With reduced chemical models, CFD simulations are feasible. Yet, CFD is too demanding for parametric studies and even more so for systems' tasks, such as optimization and control. There is an obvious need to reduce the transport models as well. A question raised then is what is the minimal reduced reactor model for microscales (see below).

Model reduction of either chemistry or transport models can be achieved using various techniques, such as asymptotics, dimensional analysis, perturbation, sensitivity and principal component analyses, proper orthogonal decomposition, computational singular perturbation, neural networks, surface response, and low-dimensional manifold method. This subject is fairly rich and not further covered herein.

The overall approach is reminiscent of kids' playing. Models are first constructed from building blocks (like Lego toys), are then refined, and eventually "demolished" to their backbones. A major difference from kids' toys is that refinement and demolition are very systematic and selective and are information-based.

10.4.2
Methodological Steps

In recent published work, we demonstrated this overall approach theoretically and experimentally for individual microchemical components, namely (i) ammonia decomposition microreactors to produce hydrogen for PEM fuel cells (see Figure 10.7) and (ii) catalytic microcombustors (ceramic ones in Figure 10.3 or tunable, stainless steel framework ones in Figure 10.4) needed to drive endothermic reactions, such as ammonia cracking. Important steps can be outlined as follows:

1. Develop detailed surface reaction mechanisms for different processes, including the oxidation of H_2, CO, CH_4, and oxygenates, the water-gas shift, the preferential

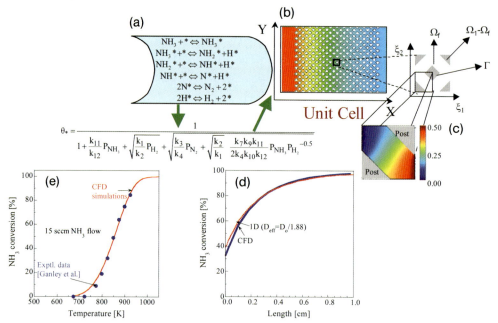

Figure 10.7 Methodological steps from post microreactor internal microstructure optimization. (a) Microkinetic and reduced model [72, 85]; (b) 2D CFD simulation of a post microreactor in NH_3 cracking producing H_2 [86]; (c) simulation of a single post to compute the concentration field and then the effective diffusivity, D_{eff}, as a function of geometric features [70]; (d) comparison of CFD simulation (CPU ~ 1 week) and 1D diffusion, convection, and reaction model run in Matlab (CPU ~ seconds) [70]; (e) comparison of model to data [86].

oxidation (PROX) of CO on noble metals, and the NH_3 decomposition on Ru [72–78].

2. Create reduced mechanisms, over a desirable operation regime, using sensitivity analysis, reaction path analysis, principal component analysis, and asymptotics [25, 79, 80].

3. Incorporate the reduced reaction mechanisms in CFD to account for the coupling of flow, transport, and chemistry [24, 45, 46].

4. Develop effective transport models, using homogenization theory, which account for geometric features (e.g. post density, shape, size) of microchemical systems [70] to enable rigorous optimization of internal microreactor features, for example posts, (see Figure 10.7).

5. Develop transport correlations, using CFD simulations, which are valid over a range of conditions [36].

6. Employ correlations and reduced chemistry in pseudo-2D models (effectively 1D models with lumped transverse heat and mass transfer) to carry out parametric studies [81, 82]. An example is shown in Figure 10.8, where ignition

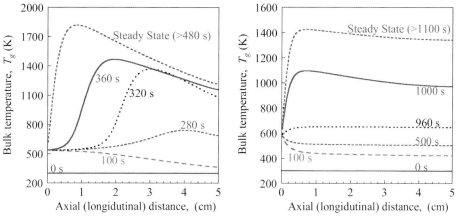

Figure 10.8 Transients in the combustion of propane/air mixtures in a catalytic microburner with Pt/Al$_2$O$_3$ with heat losses during start-up with preheating of reactants slightly above the ignition temperature. From left to right the wall conductivity varies from 2 to 200 W m^{-1} K^{-1}, changing ignition from back-end to front-end, respectively. Note the slow start-up when preheating (near the ignition temperature) is used and the importance of wall conductivity in dictating where ignition starts and the start-up time. Data taken from [35].

in a catalytic microburner is studied [30]. Simulation results make clear that heat losses, which are unavoidable at small scales, and materials of construction of microreactors are key design parameters affecting transients. Such key parameters are typically unaccounted for at large scales and underscore the importance of accounting for transient operation of microscale devices.

7. Carry out single microreactor optimization using superstructure optimization techniques [69]. Such models provide insights into integrated microsystems. An example on determining the optimal temperature profile, through a network of reactors, for the water-gas shift chemistry is shown in Figure 10.9. Substantial reduction in CO levels can be achieved using a number of microreactors in series, each at a different temperature, instead of employing a single microreactor.

10.4.3
Minimal Microreaction Engineering Models

An outcome from these simulations is that an overall materials and energy balance approach is valuable since it could provide some of the limits of operation within a factor of two (e.g. how to balance flow rates between various streams) [83]; at the same time, it cannot predict hot spots and extinction limits caused by high flow rates (e.g. when the flow rate of the endothermic channel becomes large, it removes much heat causing the exothermic reaction channel to die). A continuously stirred tank reactor (CSTR) model can provide additional information (e.g. mainly extinction limits) but is not necessary much more accurate than the overall materials

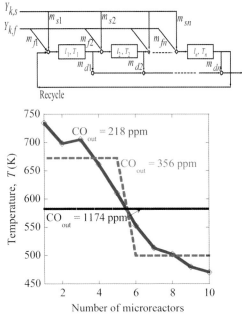

Figure 10.9 (a) Schematic of a general reactor network superstructure consisting of *n* reactors in series. (b) Application of the superstructure optimization to the water-gas shift reaction on Pt. Staging in temperature, steam, and removal of side streams was considered. In this case temperature profiling is most important (due to equilibrium limitations); a six-fold reduction in CO emissions (solid curve) can be accomplished, in comparison to a single isothermal microreactor, via connecting multiple microreactors in series operating at different temperatures in an almost linear temperature gradient. Taken from [74].

and energy balance approach. One-dimensional (heterogeneous) (also termed pseudo-two dimensional models), which describe longitudinal transport in detail and transverse transport with heat and mass transfer correlations, are the bare minimum models that capture all the important characteristics of multifunctional microchemical devices, for example the effect of wall materials, hot spots, species and temperature gradients, extinction and blowout limits, and so on [81, 84]. However, full 2D and 3D CFD simulations are needed for detailed design. These results, compounded with the need for modeling of transients, underscore a departure from the traditional systems' approach of synthesizing and optimizing flow sheets and the difficulty in optimization and control brought about from the inherent distributed nature of microchemical systems.

10.4.4
Catalyst Design

The quest for alternative, more active and selective catalysts, which are more stable under transient conditions and safe for portable and distributed power, is an ongoing process. High throughput experiments facilitate catalyst discovery but create massive databases of data that are difficult to screen and capitalize on for further materials design. Informatics tools can be valuable in creating correlations between performance and suitable descriptors, but may give false alarms and may not extrapolate or interpolate well due to nonlinearities. Furthermore, determination of suitable descriptors is more of an art at this point. Fundamental multiscale modeling can also play a role in catalyst design. This can be achieved by providing insights into the process and catalyst and using fundamental models that can be extrapolated reliably. Eventually, one may have to rely on a hybrid approach, one that bridges the gap between the empiricism of informatics/high throughput experimentation and the tremendous computational overhead but sophistication of a first principles, multiscale modeling approach [69]. Overall, this area is still in its embryonic stages.

10.5
Conclusions

Microreactors offer an opportunity for portable and distributed power generation where small to moderate power through puts are needed. This opportunity stems from their process intensification, inherent safety, compactness, and ease of scaling out. At the same time, miniaturization creates multiple challenges, summarized in Table 10.1. These challenges can be overcome by new reactor designs, proper integration of various units, and new systems (design and control) approaches. In this exercise, multiscale modeling and associated reduced models can assist the developers in both reactor and catalyst design.

Acknowledgements

This work has been funded through the Army Research Laboratory Composite Materials Research program at the University of Delaware Center for Composite Materials, the Army Research Office under contract DAAD19-01-0582, and the National Science Foundation CBET-0729714 program. Any findings, opinions, and conclusions or recommendations expressed here are those of the authors and do not necessarily reflect the views of the Army Research Laboratory or the Army Research Office. The author would also like to acknowledge undergraduate and graduate students and postdoctoral fellows whose work has contributed to this chapter.

References

1 Srinivasan, R., Hsing, I.M., Berger, P.E., Jensen, K.F., Firebaugh, S.L., Schmidt, M.A., Harold, M.P., Lerou, J.J. and Ryley, J.F. (1997) Micromachined reactors for catalytic partial oxidation reactions. *AIChE Journal*, **43**, 3059–69.

2 Lerou, J.J. and Ng, K.M. (1996) Chemical reaction engineering: a multiscale

approach to a multiobjective task. *Chemical Engineering Science*, **51**, 1595–614.

3 Microchemical Technology for Chemical and Biological Microreactors (1995) Papers of the Workshop on Microsystem Technology. Mainz, Feb. 20–21: Dechema Monographs, Vol. 132.

4 Jensen, K.F. (2005) Silicon-based microreactors. *ACS Symposium Series*, **914**, 2–22.

5 Mitchell, C.M., Kim, D.-P. and Kenis, P.J.A. (2006) Ceramic microreactor for on-site hydrogen production. *Journal of Catalysis*, **241**, 235–42.

6 Norton, D.G., Wetzel, E.R. and Vlachos, D.G. (2004) Fabrication of single-channel catalytic microburners: effect of confinement on the oxidation of hydrogen/air mixtures. *Industrial and Engineering Chemistry Research*, **43**, 4833–40.

7 Younes-Metzler, O., Svagin, J., Jensen, S., Christensen, C.H., Hansen, O. and Quaade, U. (2005) Microfabricated high-temperature reactor for catalytic partial oxidation of methane, *Applied Catalysis A: General*, **284**, 5–10.

8 Gavriilidis, A., Angeli, P., Cao, E., Yeong, K. and Wan, Y. (2002) Technology and applications of microengineered reactors. *Transactions of the Institution of Chemical Engineers*, **80**, 3–30.

9 Jensen, K.F. (2001) Microreaction engineering – is small better? *Chemical Engineering Science*, **56**, 293–303.

10 Kothare, M.V. (2006) Dynamics and control of integrated microchemical systems with application to micro-scale fuel processing. *Computers and Chemical Engineering*, **30**, 1725–34.

11 Kolb, G. and Hessel, V. (2004) Review: Micro-structured reactors for gas phase reactions. *Chemistry-Engineering Journal*, **98**, 1–38.

12 Keil, F.J. (2004) Catalytic reactions and reactors. *Chemical Engineering Science*, **59**, 5473–8.

13 Jähnisch, K., Hessel, V., Löwe, H. and Baerns, M. (2004) Chemistry in microstructured reactors, *Angewandte Chemie – International Edition*, **43**, 406–46.

14 Norton, D.G., Deshmukh, S.R., Wetzel, E.D. and Vlachos, D.G. (2005) Downsizing chemical processes for portable hydrogen production, in *Microreactor Technology and Process Intensification* (eds Y. Wang and J.D. Holladay), ACS, New York, **914**, pp. 179–93.

15 Mitsos, A., Palou-Rivera, I. and Barton, P.I. (2004) Alternatives for micropower generation processes. *Industrial and Engineering Chemistry Research*, **43**, 74–84.

16 Miesse, C.M., Masel, R.I., Jensen, C.D., Shannon, M.A. and Short, M. (2004) Submillimeter-scale combustion. *AIChE Journal*, **50**, 3206–14.

17 DOE17. (2004) On-board fuel processing go/no-go decision: DOE decision team committee report, http://www1.eere.energy.gov/hydrogenandfuelcells/pdfs/committee_report.pdf (accessed November 17, 2006)

18 Hoogers, G. (ed.) (2003) *Fuel Cell Technology*, CRC Press, Handbook.

19 Tonkovich, A.Y., Yang, B., Perry, S.T., Fitzgerald, S.P. and Wang, Y. (2007) From seconds to milliseconds to microseconds through tailored microchannel reactor design of a steam methane reformer. *Catalysis Today*, **120**, 21–9.

20 Hickman, D.A. and Schmidt, L.D. (1993) Production of synthesis gas by direct catalytic oxidation of methane. *Science*, **259**, 343–6.

21 Goetsch, D.A. and Schmidt, L.D. (1996) Microsecond catalytic partial oxidation of alkanes. *Science*, **271**, 1560–2.

22 Beretta, A., Ranzi, E. and Forzatti, P. (2001) Production of olefins via oxidative dehydrogenation of light paraffins at short contact times. *Catalysis Today*, **64**, 103–11.

23 Ganley, J.C., Seebauer, E.G. and Masel, R.I. (2004) Porous anodic alumina microreactors for production of hydrogen from ammonia. *AIChE Journal*, **50**, 829–34.

24 Deshmukh, S.R., Mhadeshwar, A.B. and Vlachos, D.G. (2004) Microreactor modeling for hydrogen production from ammonia decomposition on ruthenium. *Industrial and Engineering Chemistry Research*, **43**, 2986–99.

25 Deshmukh, S.R., Mhadeshwar, A.B., Lebedeva, M.I. and Vlachos, D.G. (2004) From density functional theory to microchemical device homogenization:

model prediction of hydrogen production for portable fuel cells. *International Journal for Multiscale Computational Engineering*, **2**, 221–38.

26 Federici, J.A., Norton, D.G., Brüggemann, T., Voit, K.W., Wetzel, E.D. and Vlachos, D.G. (2006) Catalytic microcombustors with integrated thermoelectric elements for portable power production. *Journal of Power Sources*, **161**, 1469–78.

27 Deshmukh, S.R. and Vlachos, D.G. (2005) Novel micromixers driven by flow instabilities: application to post-reactors. *AIChE Journal*, **51**, 3193–204.

28 Bunluesin, T., Gorte, R.J. and Graham, G.W. (1998) Studies of the water-gas shift reaction on ceria-supported Pt, Pd, and Rh: implications for oxygen-storage properties. *Applied Catalysis B: Environmental*, **15**, 107–14.

29 Norton, D.G., Wetzel, E.D. and Vlachos, D.G. (2006) Thermal management in catalytic microreactors. *Industrial and Engineering Chemistry Research*, **45**, 76–84.

30 Kaisare, N., Stefanidis, G.D. and Vlachos, D.G. (2008) Comparison of ignition strategies for catalytic microburners. *Proc. Comb. Inst*, in press.

31 Barton, P.I., Mitsos, A. and Chachuat, B. (2005) Optimal start-up of micro power generation processes, in *Computer Aided Chemical Engineering*, Vol. 20B (eds C. Puigjaner and A. Espuna), Elsevier, Barcelona, Spain, pp. 1093–8, 29th May–1st June, ESCAPE 15B.

32 Chachuat, B., Mitsos, A. and Barton, P.I. (2005) Optimal start-up of micro power generation processes employing fuel cells, in *AIChE Annual Meeting, Cincinnati, OH, 30th October–34th November.*

33 Wheeler, C., Jhalani, A., Klein, E.J., Tummala, S. and Schmidt, L.D. (2004) The water-gas shift reaction at short contact times. *Journal of Catalysis*, **223**, 191–9.

34 Hilaire, S., Wang, X., Luo, T., Gorte, R.J. and Wagner, J. (2001) A comparative study of water-gas shift reaction over ceria supported metallic catalysts. *Applied Catalysis A: General*, **215**, 271–8.

35 Fu, Q., Saltsburg, H. and Flytzani-Stephanopoulos, M. (2003) Active non-metallic Au and Pt species on ceria-based

water-gas shift catalysts. *Science*, **301**, 935–8.

36 Kaisare, N., Stefanidis, G.D. and Vlachos, D.G. (2007) Transport phenomena in microscale reacting flows, in *Micro Process Engineering: Fundamentals, Operations, and Catalysts*, Vol. 1 (eds A. Renken and Y. Wang), Wiley-VCH Verlag GmbH (accepted).

37 Norton, D.G. and Vlachos, D.G. (2003) Combustion characteristics and flame stability at the microscale: a CFD study of premixed methane/air mixtures, *Chemical Engineering Science*, **58**, 4871–82.

38 Norton, D.G. and Vlachos, D.G. (2004) A CFD study for propane/air microflame stability. *Combustion and Flame*, **138**, 97–107.

39 Kolb, G., Cominos, V., Hofmann, C., Pennemann, H., Schurer, J., Tiemann, D., Wichert, M., Zapf, R., Hessel, V. and Lowe, H. (2005) Integrated microstructured fuel processors for fuel cell applications. *Chemical Engineering Research and Design*, **83**, 626–33.

40 Seris, E.l.E., Abramowitz, G., Johnston, A.M. and Haynes, B.S. (2005) Demonstration plant for distributed production of hydrogen from steam reforming of methane. *Chemical Engineering Research and Design*, **83**, 619–25.

41 Zanfir, M. and Gavriilidis, A. (2004) Influence of flow arrangement in catalytic plate reactors for methane steam reforming. *Chemical Engineering Research and Design*, **82**, 252–8.

42 Venkataraman, K., Redenius, J.M. and Schmidt, L.D. (2002) Millisecond catalytic wall reactors: dehydrogenation of ethane. *Chemical Engineering Science*, **57**, 2335–43.

43 Kolios, G., Glokler, B., Gritsch, A., Morillo, A. and Eigenberger, G. (2005) Heat-integrated reactor concepts for hydrogen production by methane steam reforming. *Fuel Cells*, **5**, 52–65.

44 Venkataraman, K., Wanat, E.C. and Schmidt, L.D. (2003) Steam reforming of methane and water-gas shift in catalytic wall reactors. *AIChE Journal*, **49**, 1277–84.

45 Deshmukh, S.R. and Vlachos, D.G. (2005) CFD simulations of coupled, counter-current combustor/reformer microdevices for hydrogen production. *Industrial and Engineering Chemistry Research*, **44**, 4982–92.

46 Deshmukh, S.R. and Vlachos, D.G. (2005) Effect of flow configuration on the operation of coupled combustor/ reformer microdevices for hydrogen production. *Chemical Engineering Science*, **60**, 5718–28.

47 Lloyd, S.A. and Weinberg, F.J. (1974) A burner for mixtures of very low heat content. *Nature*, **251**, 47–9.

48 Jones, A.R., Lloyd, S.A. and Weinberg, F.J. (1978) Combustion in heat exchangers, *Proceedings of the Royal Society of London. Series A*, **360**, 97–115.

49 Ronney, P.D. (2003) Analysis of non-adiabatic heat-recirculating combustors. *Combustion and Flame*, **135**, 421–39.

50 Vican, J., Gajdeczko, B.F., Dryer, F.L., Milius, D.L., Aksay, I.A. and Yetter, R.A. (2003) Development of a microreactor as a thermal source for microelectromechanical systems power generation. *Proceedings of the Combustion Institute*, **29**, 909–16.

51 Chen, M. and Buckmaster, J. (2004) Modelling of combustion and heat transfer in "Swiss roll" micro-scale combustors. *Combustion Theory and Modelling*, **8**, 701–20.

52 Ahn, J.M., Eastwood, C., Sitzki, L. and Ronney, P.D. (2005) Gas-phase and catalytic combustion in heat-recirculating burners. *Proceedings of the Combustion Institute*, **30**, 2463–72.

53 Ju, Y. and Choi, C.W. (2003) An analysis of sub-limit flame dynamics using opposite propagating flames in mesoscale channels. *Combustion and Flame*, **133**, 483–93.

54 Sitzki, L., Borer, K., Schuster, E., Ronney, P.D. and Wussow, S. (2001) Combustion in microscale heat-recirculating burners, in *The Third Asia-Pacific Conference on Combustion, Seoul, Korea*.

55 Kim, N.I., Kato, S., Kataoka, T., Yokomori, T., Maruyama, S., Fujimori, T. and Maruta, K. (2005) Flame stabilization and emission of small Swiss-roll combustors as heaters. *Combustion and Flame*, **141**, 229–40.

56 Kim, N., Aizumi, S., Yokomori, T., Kato, S., Fujimori, T. and Maruta, K. (2007) Development and scale effects of small

swiss-roll combustors. *Proceedings of the Combustion Institute*, **31**, 3243–50.

57 Friedle, U. and Veser, G. (1999) A counter-current heat-exchange reactor for high temperature partial oxidation reactions. *Chemical Engineering Science*, **54**, 1325–32.

58 Matros, Y.S. and Bunimovich, G.A. (1996) Reverse-flow operation in fixed bed catalytic reactors. *Catalysis Reviews-Science and Engineering*, **38**, 1–68.

59 Liu, T., Gepert, V. and Veser, G. (2005) Process intensification through heat-integrated reactors for high-temperature millisecond contact-time catalysis. *Chemical Engineering Research & Design*, **63**, 611–18.

60 Neumann, D. and Veser, G. (2005) Catalytic partial oxidation of methane in a high-temperature reverse-flow Reactor. *AIChE Journal*, **51**, 210–23.

61 Kulkarni, M.S. and Dudukovic, M.P. (1996) A bidirectional fixed-bed reactor for coupling of exothermic and endothermic reactions. *AIChE Journal*, **42**, 2897–910.

62 Kulkarni, M.S. and Dudukovic, M.P. (1998) Periodic operation of asymmetric bidirectional fixed-bed reactors with temperature limitations. *Industrial and Engineering Chemistry Research*, **37**, 770–81.

63 Kaisare, N.S. and Vlachos, D.G. (2007) Extending the region of stable homogeneous micro-combustion through forced unsteady operation. *Proceedings of the Combustion Institute*, **31**, 3293–300.

64 de Smet, C.R.H., de Croon, M.H.J.M., Berger, R.J., Marin, G.B. and Schouten, J.C. (2001) Design of adiabatic fixed-bed reactors for the partial oxidation of methane to synthesis gas. Application to production of methanol and hydrogen-for-fuel-cells. *Chemical Engineering Science*, **56**, 4849–61.

65 Vlachos, D.G. (2005) A review of multiscale analysis: Examples from systems biology, materials engineering, and other fluid-surface interacting systems. *Advanced Chemical Engineering*, **30**, 1–61, invited.

66 Raimondeau, S., Aghalayam, P., Vlachos, D.G. and Katsoulakis, M. (2001) Bridging the gap of multiple scales: From microscopic, to mesoscopic, to macroscopic models, in *Foundations of Molecular Modeling and Simulation, AIChE*

Symposium Series No. 325, 97, Keystone, CO, USA, pp. 155–8.

67 Raimondeau, S. and Vlachos, D.G. (2002) Recent developments on multiscale, hierarchical modeling of chemical reactors. *Chemistry-Engineering Journal*, **90**, 3–23.

68 Chatterjee, A. and Vlachos, D.G. (2007) Systems tasks in nanotechnology via hierarchical multiscale modeling: nanopattern formation in heteroepitaxy. *Chemical Engineering Science*, **62**, 4852–63.

69 Vlachos, D.G., Mhadeshwar, A.B. and Kaisare, N. (2006) Hierarchical multiscale model-based design of experiments, catalysts, and reactors for fuel processing. *Computers and Chemical Engineering*, **30**, 1712–24.

70 Deshmukh, S.R., Mhadeshwar, A.B., Lebedeva, M.I. and Vlachos, D.G. (2004) From density functional theory to microchemical device homogenization: model prediction of hydrogen production for portable fuel cells. *International Journal for Multiscale Computational Engineering*, **2**, 221–38.

71 Raimondeau, S. and Vlachos, D.G. (2000) Low-dimensional approximations of multiscale epitaxial growth models for microstructure control of materials. *Journal of Computational Physics*, **160**, 564–76.

72 Mhadeshwar, A.B., Kitchin, J.R., Barteau, M.A. and Vlachos, D.G. (2004) The role of adsorbate-adsorbate interactions in the rate controlling step and most abundant reaction intermediate of NH_3 decomposition on Ru. *Catalysis Letters*, **96**, 13–22.

73 Mhadeshwar, A.B., Aghalayam, P., Papavassiliou, V. and Vlachos, D.G. (2003) Surface reaction mechanism development for platinum catalyzed oxidation of methane. *Proceedings of the Combustion Institute*, **29**, 997–1004.

74 Mhadeshwar, A.B. and Vlachos, D.G. (2004) Microkinetic modeling for water-promoted CO oxidation, water-gas shift, and preferential oxidation of CO on Pt. *The Journal of Physical Chemistry B*, **108**, 15246–58.

75 Mhadeshwar, A.B. and Vlachos, D.G. (2005) A thermodynamically consistent surface reaction mechanism for the CO oxidation on Pt. *Combustion and Flame*, **142**, 289–98.

76 Mhadeshwar, A.B. and Vlachos, D.G. (2005) Hierarchical, multiscale surface reaction mechanism development: CO and H2 oxidation, water-gas shift, and preferential oxidation of CO on Rh. *Journal of Catalysis*, **234**, 48–63.

77 Mhadeshwar, A.B. and Vlachos, D.G. (2005) Hierarchical multiscale mechanism development for methane partial oxidation and reforming, and for thermal decomposition of oxygenates on Rh. *The Journal of Physical Chemistry B*, **109**, 16819–35.

78 Mhadeshwar, A.B. and Vlachos, D.G. (2007) A catalytic reaction mechanism for methane partial oxidation at short contact times, reforming, and combustion, and for oxygenate decomposition and oxidation on Pt. *Industrial and Engineering Chemistry Research*, **46**, 5310–24.

79 Deshmukh, S. and Vlachos, D.G. (2007) A reduced mechanism for methane and one-step rate expressions for fuel-lean catalytic combustion of small alkanes on noble metals. *Combustion and Flame*, **149**, 366–83.

80 Mhadeshwar, A.B. and Vlachos, D.G. (2005) Is the water-gas shift reaction on Pt simple? Computer-aided microkinetic model reduction, lumped rate expression, and rate-determining step. *Catalysis Today*, **105**, 162–72.

81 Kaisare, N.S. and Vlachos, D.G. (2007) Optimal reactor dimensions for homogeneous combustion in small channels. *Catalysis Today*, **120**, 96–106.

82 Kaisare, N.S., Deshmukh, S.R. and Vlachos, D.G. (2008) Stability and performance of catalytic microreactors: Simulations of propane catalytic combustion on Pt. *Chemical Engineering Science*, **63** (4), 1098–116.

83 Stefanidis, G.D., Kaisare, N., Maestri, M. and Vlachos, D.G. (2008) Methane steam reforming at microscales: Operation strategies for variable power output at millisecond contact times. *AIChE Journal*, in press.

84 Kaisare, N.S. and Vlachos, D.G. (2007) Extending the region of stable homogeneous micro-combustion through forced unsteady operation. *Proceedings of the Combustion Institute*, **31**, 3293–300.

85 Mhadeshwar, A.B., Kitchin, J.R., Barteau, M.A. and Vlachos, D.G. (2004) The role of adsorbate-adsorbate interactions in the rate controlling step and the most abundant reaction intermediate of NH_3 decomposition on Ru. *Catalysis Letters,* **96**, 13–22.

86 Deshmukh, S.R., Mhadeshwar, A.B. and Vlachos, D.G. (2004) Microreactor modeling for hydrogen production from ammonia decomposition on ruthenium. *Industrial and Engineering Chemistry Research,* **43**, 2986–99.

11
Optimal Design and Steady-State Operation

Benoît Chachuat

11.1
Introduction

The focus of this chapter is on describing procedures for the optimal design and operation of microfabricated power generation processes. In contrast to the problem of selection of alternatives, as described earlier in Chapter 8, a fixed process structure is considered throughout this chapter.

In many applications, the power demand remains essentially constant during the operation, with rapid changeovers and, therefore, steady-state operation must be considered thoroughly. For a given power demand, or a power demand varying in a specified range, the design and steady-state operation problem is to determine values of the design variables (e.g. sizes of the individual components such as fuel processing reactor and fuel cell) as well as operational variables (e.g. fuel flow rates and operating temperature) so as to maximize its (steady-state) performance, in the light of safety and reliability, as well as other considerations.

In other applications, the power demand may change rapidly and the devices may be operated periodically, with frequent start-ups and shut-downs. In this case, special attention must be paid to the dynamics of these processes, in addition to their steady-state operation. Ideally, one would like to optimize the design and operation over an entire operation cycle, including start-up, steady-state operation, and shut-down. However, because the duration and power demand profile of a mission is rarely known in advance, such an optimization is hardly tractable. Instead, the start-up phase can be optimized separately from the subsequent steady-state operation. For start-up purpose, the devices will most likely be coupled with a small battery or ultra-capacitor, whose role will be to ensure that the power demand is met when the fuel cell is unavailable or can only satisfy part of the demand, and to provide the energy needed to heat the fuel cell stack up to a temperature at which chemical and electrochemical reactions are fast enough. That is, the objective of the optimal start-up problem is to bring the fuel cell to a desired operating point (steady state), in a manner that minimizes the mass of fuel and battery required, while meeting the nominal power demand at all times, along with operational restrictions and safety requirements.

Microfabricated Power Generation Devices. Edited by Alexander Mitsos and Paul I. Barton
Copyright © 2009 WILEY-VCH Verlag GmbH & Co. KGaA, Weinheim
ISBN: 978-3-527-32081-3

At the microscale, a different paradigm to the unit-operations paradigm of macroscale process design and operation is necessary. The reason for this is that the different units constituting a microfabricated system are tightly spatially integrated and can no longer be considered to operate independently from each other. Accordingly, operational decisions must be taken at an early stage of development, together with design decisions. For example, increasing the operating temperature of a microfabricated reactor increases the heat losses per unit surface area, but because the reaction rates are also enhanced, the volume needed to achieve a given conversion is reduced. In the case where the latter effect dominates, one then obtains the counterintuitive result that increasing the operating temperature lowers the overall heat losses for the system. In other words, it is of paramount importance to determine the operation policy simultaneously with the design and sizing of the units.

Because the underlying physicochemical phenomena are complicated and intrinsically coupled, one cannot rely only on engineering intuition to find the optimal design and operation. The use of mathematical models along with systematic optimization methods based on mathematical programming is clearly warranted. Because optimization algorithms may require hundreds, or even thousands, of iterations to converge, fast, reliable and robust solution of the models is needed. These considerations proscribe the use of computationally expensive models based on computational fluid dynamics (CFD), as described earlier in Chapter 10. System-level models, as presented in Chapter 8, are also not adequate, because they cannot represent the couplings between design and operational decisions.

In this chapter, the focus is on the so-called *intermediate-fidelity models*. These models are spatially distributed, rely on validated kinetic expressions, and allow optimization of unit sizes and operation for a given process structure without the need to specify a detailed geometry. This level of modeling detail is especially useful for technologies with a demonstrated proof of principle.

In the remainder of this chapter, first the formulation of intermediate-fidelity models is discussed (Section 11.2). Then, the problem of optimal design and operation is addressed, both in the case of a nominal power demand (Section 11.3) and of a variable power demand (Section 11.4). The emphasis is placed on explaining the problem formulations and illustrating the benefit of these approaches through a number of case studies; the concrete techniques used to solve these problems are not explained.

11.2
Models for Optimal Design and Operation

The devices that are considered for man-portable power generation involve complex geometries, multiple scales, time dependence, and parametric uncertainty. Therefore, with current computational capabilities and available algorithms, the optimal design and operation problem cannot be solved in a single step. The system-level approach described earlier in Chapter 8 (see also [1, 2]), allows one to compare many different technology alternatives for man-portable power generation based on a process superstructure, including thousands of different designs. Although it allows one to identify conditions under which the technologies considered are a promising alternative to batteries, such an approach has currently the limitation that it requires values to be set for some modeling parameters that, in principle, can be calculated, for example fuel cell efficiency, conversions, etc. Furthermore, it cannot predict optimal values for several key operating parameters, such as the operating temperature and the fuel/air ratios, nor does it consider emissions of trace components, such as carbon monoxide, ammonia, or nitric oxide.

At the other extreme, the development of CFD models is a very versatile tool for detailed analysis (see Chapter 10; compare [3, 4]) as well as justification of modeling assumptions [5]. However, CFD requires comprehensive knowledge of the detailed geometric design, and an extensive modeling effort. Moreover, it can be very computationally expensive. Consequently, the applicability of CFD to optimization of the design and operation is currently rather limited.

The foregoing considerations motivate the development of models of an intermediate detail level to study the optimal sizing of units and optimal operation. In order for the couplings between the design and operational decisions to be accurately represented, these models rely on first principles and include detailed kinetic mechanisms. Spatial dependence is considered whenever necessary but, unlike CFD, a fully defined geometry is not required. Rather, a minimal number of design parameters need to be specified, such as the volume and the surface-area-to-volume-ratio of the units. Intermediate-fidelity models were initially developed in the context of steady-state operation [6, 7], which results in models that are comprised of mixed sets of differential and algebraic equations (DAEs). In the case of transient operation, integration through time must be considered in addition to the spatial distribution, which leads to models that consist of partial differential algebraic equations (PDAEs) [8, 9]. Only steady-state aspects will be discussed in this chapter. The design, operation, and control issues associated with the transient operation of man-portable power generation processes are addressed in the subsequent Chapters 12 and 13.

Obviously, the nature of intermediate-fidelity models is dependent on the class of devices considered. Valid approximations can be established by detailed modeling, scaling analysis, and experimental evidence. Typical simplifications, which are valid for a specific class of devices only, are given for illustrative purpose next. For high-temperature systems with maximal characteristic dimensions of the order of millimeters and high thermal conductivity, a good approximation is to assume a spatially uniform temperature. In practice, the use of silicon, together with the presence of a catalyst support, ensures such a high thermal conductivity [5]. However, the approximation of uniform temperature may not be adequate for all microstructured reactors, for example in the case of combustion [10]. Besides temperature, a one-dimensional distribution of the species balance appears to be an adequate approximation for reactors that are made of thin tubes, for example for tubes with diameters of the order of $100\,\mu m$ [11].

The formulation of an intermediate-fidelity model for a simple man-portable power generation process is presented in the following case study.

11.2.1
Case Study

A micro power generation process that consists of a fuel processing reaction, a solid-oxide fuel cell (SOFC) and two burners, microfabricated into a single silicon stack fed with ammonia and butane fuels is studied. This technology alternative satisfies three important criteria, namely: (i) the potential for high performance, which is ascertained from system-level considerations; (ii) a demonstrated proof of concept, and (iii) the availability of validated chemical kinetics. Here, the SOFC is chosen because of the long-term promise for fuel flexibility [12]. Although butane partial oxidation combined with a SOFC is expected to have higher energy density, the corresponding chemistry has not yet been sufficiently demonstrated in microreactors. On the other hand, ammonia decomposition has been successfully performed with conversions exceeding 90% [3, 13]. This motivated the choice of ammonia as primary fuel, despite the limitation of potential applications due to its toxicity. Moreover, butane is chosen as secondary fuel because of its high energy density, which makes it a suitable heat source.

The process is arranged in two lines (Figure 11.1). The ammonia line begins with a reactor for the catalytic decomposition of ammonia (NH_3) to produce hydrogen (H_2). H_2 is fed to the SOFC anode while air is fed to the SOFC cathode, and electrical power is produced from the electrochemical reaction. The effluents from the anode and cathode compartments of the SOFC are then passed to Burner I for catalytic oxidation of residual H_2 and NH_3. The butane line consists of Burner II, which is fed with a mixture of butane (C_4H_{10}) and air in order to produce heat from catalytic oxidation.

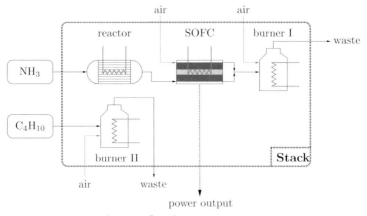

Figure 11.1 Conceptual process flow-sheet.

A summary of the main assumptions used to formulate an intermediate-fidelity model for this process is given hereafter. For a complete statement of the model equations, please refer to [6, 7].

11.2.1.1 Intermediate-Fidelity Model

The main assumptions used in the model formulation are the following:

1. The pressure inside the stack is uniform and equals the atmospheric pressure. This assumption requires that the pressure drop along the gas channels remains relatively small, which is typically the case for characteristic lengths as small as 50 µm in the radial direction and a few cm in the axial direction.

2. The gas phase is ideal. This assumption appears to be reasonable when the stack operates at elevated temperature and atmospheric pressure.

3. The stack operates at a spatially uniform temperature. This assumption requires that heat transfer is fast enough, which is typically the case at the microscale for silicon-based reactors due to the high thermal conductivity of silicon and the small length scales involved.

4. The molar fluxes in the gas channels of the four units are convective in the flow direction (no axial diffusion) and radial gradients are neglected. This last assumption asserts that microfabricated units such as reactors or fuel cells can be approximated well by an idealized model using 1-D distributed models. Neglecting axial diffusion is motivated by the fact that the Peclet number in the axial direction is relatively large; neglecting radial gradients, on the other hand, appears to be a legitimate assumption since the characteristic time for diffusion is generally smaller than the characteristic time for reaction.

Based on the assumptions 1–4, the mass and species equations for each unit can be written as that of a plug-flow reactor (PFR) at steady-state [14]:

$$\frac{dF}{d\eta} = V \sum_{j=1}^{n_r} \sum_{i \in I_s} v_{i,j} r_j \tag{11.1}$$

$$\frac{dz_i}{d\eta} = \frac{V}{F} \sum_{j=1}^{n_r} \left[v_{i,j} - z_i \sum_{k \in I_s} v_{k,j} \right] r_j, \quad i \in I_s \tag{11.2}$$

In these equations, $\eta \in [0, 1]$ stands for the dimensionless coordinate in the flow direction; V is the volume of the unit; T is the operating temperature; F is the total molar flow; z_i is the mole fraction of species i; r_j is the rate of reaction j; $v_{i,j}$ is the stoichiometric coefficient for species i in reaction j; $I_s := \{NH_3, C_4H_{10}, H_2, H_2O, O_2, N_2, NO, CO_2\}$; and n_r is the number of reactions. In addition to the differential Equations 11.1 and 11.2, the unit models contain algebraic equations to calculate the reaction rates r_j as a function of the molar fractions z_i, $I \in I_s$, as well as operational variables such as the operating temperature T or the fuel cell voltage U; some of these algebraic equations are explicit (e.g. the reaction rates), and some are implicit (e.g. the Butler–Volmer equations). Overall, each unit is thus described by a set of DAEs.

By putting the DAEs describing each unit together, one obtains a global model for the fuel cell stack in the form of multistage DAEs, wherein each stage corresponds to a given unit. The solutions of the equations for each stage in the ammonia line are coupled: the mole fractions at the exit of the reactor affect the initial conditions for the DAEs describing the SOFC; in turn, the mole fractions at the exit of the SOFC affect the initial conditions for the DAEs describing Burner I.

The steady-state operation and decision variables in each unit of the fuel cell stack can be summarized as follows:

Ammonia Line:

1. **Reactor:** NH_3 is endothermically decomposed using a catalyst according to the overall reaction:

$$NH_3 \rightarrow \frac{3}{2}H_2 + \frac{1}{2}N_2$$

The reduced kinetic rate expressions proposed in [15] for ammonia decomposition over ruthenium (temperature dependent) can be used as a first approximation. In the reactor model, the reactor volume, V^r, is a design decision variable; the optimal NH_3 inlet flow rate, F^r_{in}, together with the operating temperature, are operational decision variables.

2. **SOFC:** The solid-oxide fuel cell consists of a cathode and an anode that are separated by a solid electrolyte. In the cathode compartment, oxygen from the air is converted to oxygen ions. The oxygen ions migrate to the anode side through the ion-conducting electrolyte. In the anode compartment, hydrogen reacts with oxygen ions to produce water. Electrons flow back to the cathode via an external circuit. The half cell reactions of the process are:

$$\text{cathode:} \frac{1}{2}O_2 + 2e^- \rightarrow O^{2-}, \quad \text{anode:} H_2 + O^{2-} \rightarrow H_2O + 2e^-$$

The open-circuit potential for the SOFC is taken as the Nernst potential. Given the operating voltage of the fuel cell, U, one can then calculate the electrical current density, j. This can be done by accounting for the irreversibilities arising from both ohmic losses through the electrolyte [16] and activation polarizations at the anode and cathode sides [17]; on the other hand, it has been shown that irreversibilities in the form of concentration overpotentials at both the anode and cathode sides can be neglected [6]. The rate of electrochemical reaction is obtained from the electrical current density by Faraday's law. Further, the electrical power generated by the fuel cell, $\mathcal{P}$, can be calculated as:

$$\mathcal{P} := \frac{A^{fc}V^{fc}}{2}U\int_0^1 j\,d\eta \tag{11.3}$$

with V^{fc} standing for the fuel cell volume, and A^{fc} being the surface-area-to-volume ratio of the anode and cathode compartments for the electrochemical reaction. In the SOFC model, the fuel cell volume is a design decision variable; the air inlet flow rate, F_{in}^{ca}, the fuel cell voltage and the operating temperature are operational decision variables.

3. **Burner I:** The effluents from the SOFC anode and cathode are passed to Burner I, possibly with an additional air stream. Catalytic oxidation of the residual H_2 and NH_3 can provide heat to balance heat losses as well as the endothermic decomposition of NH_3 in the reactor; part of the residual NH_3 is also decomposed into H_2. The overall reactions occurring in Burner I are:

$$NH_3 + \frac{5}{2}O_2 \rightarrow NO + \frac{3}{2}H_2O, \quad NH_3 + \frac{3}{2}NO \rightarrow \frac{5}{4}N_2 + \frac{3}{2}H_2O,$$

$$NH_3 \rightarrow \frac{1}{2}N_2 + \frac{3}{2}H_2, \quad H_2 + \frac{1}{2}O_2 \rightarrow H_2O$$

The kinetic rate expressions proposed in [18] (temperature dependent) can be used to model these reactions as a first approximation. At the temperatures of operation of the fuel cell (around 1000–1300 K), the unimolecular decomposition of NO is not experimentally observed, therefore it is ignored in the model. In the Burner I model, the burner volume, V^{bI}, is a design decision variable; the burner inlet air flow rate, F_{in}^{bI}, and the operating temperature are operational decision variables.

Butane Line:

4. **Burner II:** The combustion of C_4H_{10} supplies additional heat to balance the heat losses to the ambient and the endothermic decomposition of NH_3. A premixed butane/air mixture is fed to the burner, with an oxygen stoichiometry value of 1.2 so that the combustion occurs with excess air. Butane combustion is modeled as a homogeneous irreversible, one-step reaction:

$$C_4H_{10} + \frac{13}{2}O_2 \rightarrow 4CO_2 + 5H_2O$$

As a first approximation, the kinetic rate expression proposed by [19] can be used. In the burner II model, the volume of the burner, V^{bII}, is a design decision variable; the inlet flow rate of butane, F_{in}^{bII}, and the operating temperature are operational decision variables.

For simulation of the intermediate-fidelity model, in addition to all the chemical and thermodynamical properties, inlet compositions and parameters (for which the values as in [6] are used herein), one needs to specify ten degrees of freedom. Four of them correspond to design decisions: the volumes V^r, V^{fc}, V^{bI}, and V^{bII} of the reactor, fuel cell, burner I and burner II, respectively; the remaining six degrees of freedom correspond to operational decisions: the operating temperature T, the fuel cell voltage U, the ammonia inlet flow rate F_{in}^r to the reactor, the air inlet flow rates F_{in}^{ca} to the fuel cell cathode and F_{in}^{bi} to burner I, and the butane inlet flow rate F_{in}^{bII} in to burner II.

11.2.1.2 Steady-State Simulations

Simulation results are presented to illustrate the operation of the ammonia line. The following unit volumes are used: $V^r = V^{bi} = 9.6 \times 10^{-10}$ m^3 and $V^{fc} = 9.6 \times 10^{-8}$ m^3. Observe that the fuel cell volume (i.e. the sum of both anode and cathode compartments) is taken to be a hundred times larger than either the reactor or burner I volumes since the electrochemical reactions are much slower than the ammonia decomposition and hydrogen/ammonia oxidation reactions. Furthermore, the inlet flow rates of ammonia to the reactor and air to the cathode and burner I are set to: $F_{in}^r = 15$ sccm, $F_{in}^{ca} = 100$ sccm, and $F_{in}^{bi} = 0$ sccm. This latter value indicates that no additional air is fed into burner I, assuming that the fuel cell is operated at a high oxygen excess and thus enough residual oxygen remains in burner I's feed. Finally, an operating temperature $T = 1300$ K and a cell voltage $U = 0.65$ V are considered. These specifications were chosen since they correspond to a production of electrical power close to $\mathcal{P} = 1$ W.

The component mole fractions along the gas channels in the reactor, anode, cathode and burner I are presented in Figure 11.2. Based on the gas molar flow rates (not shown in Figure 11.2) and compositions, performance factors can be easily calculated for each unit of the device. It should be mentioned that these factors can be used as parameters at the system level (see Chapter 8), and the intermediate-fidelity model can thus be used as a tool to estimate their values or for validation purposes.

- The conversion of ammonia in the reactor is calculated to be $\zeta_{NH_3}^r \approx 0.987$. This value indicates that most of the ammonia is converted into hydrogen.

- The micro-SOFC operation can be monitored by considering three performance factors, namely the hydrogen conversion in the anode compartment, ζ^{an}, the air excess number in the cathode compartment, Φ^{ca}, and the fuel cell efficiency, η^{fc}. The latter factor accounts for the losses induced by the irreversibilities in the fuel cell (i.e. $\eta^{fc} = 100\%$ when no current is passed through the fuel cell). The following values are obtained for the conditions shown in Figure 11.2: $\zeta^{an} \approx 0.503$, $\Phi^{ca} \approx 3.76$, and $\eta^{fc} \approx 0.688$. These values indicate that neither the available hydrogen nor the supplied oxygen are limiting for the electrochemical reaction, but the hydrogen conversion is low, mainly due to an insufficient residence time in the anode compartment. Furthermore, the high SOFC efficiency indicates that the irreversibilities remain limited.

- The conversions of hydrogen and ammonia in burner I are calculated to be $\zeta_{H_2}^{bi} \approx 0.516$ and $\zeta_{NH_3}^{bi} \approx 0.999$. The latter value shows that the conversion of ammonia is nearly complete, whereas only approximately half of the residual hydrogen is converted from the former.

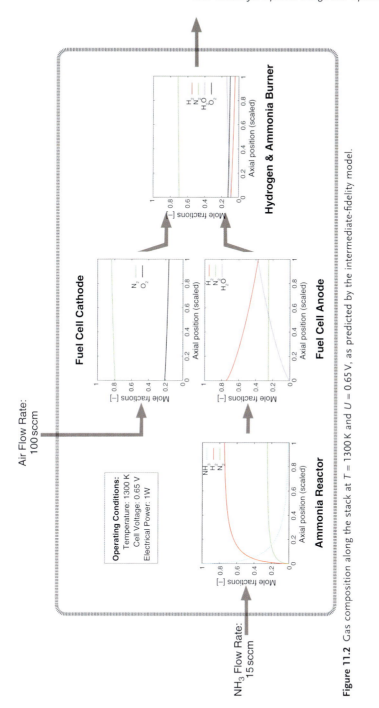

Figure 11.2 Gas composition along the stack at $T = 1300\,K$ and $U = 0.65\,V$, as predicted by the intermediate-fidelity model.

This rapid analysis reveals that the process does not operate very efficiently under these specific conditions, as a large fraction of the hydrogen produced in the reactor is discarded from the device without being used. Moreover, the residual mole fraction of nitric oxide is relatively high, of the order of 1500 ppm, hence greatly exceeding the currently allowed safe level (neither the concentration of NH_3 nor that of NO are represented in Figure 11.2 because their values are much lower than those obtained for the other components H_2, H_2O, N_2 and O_2). Accordingly, the process performance could be greatly improved with the specification of more suitable design and operational variable values. These aspects are the focus of the following two sections.

11.3
Optimal Design and Steady-State Operation for Nominal Power Demand

Intermediate-fidelity models are invaluable tools in the design of man-portable power generation systems since they provide a picture of the gas composition and velocities, potential and current density in the system for various process configurations and operating conditions. Not only can such models be used to examine the sensitivity of the system with respect to relevant design parameters and operational variables, but this sensitivity information can also be used to improve the system performance through the application of systematic optimization methods.

Qualitatively, the optimization problem for optimal design and steady-state operation of man-portable power generation processes can be stated as follows:

> "Determine the design and operational decisions that maximize the device performance, while meeting the nominal power demand, guaranteeing autothermal operation, and respecting the safety and operational constraints."

Because intermediate-fidelity models are comprised of multistage DAEs, this formulation gives rise to challenging, constrained dynamic optimization problems. Efficient methods exist for the numerical solution of such problems [20], yet their description lies beyond the scope of this chapter.

A number of remarks are in order:

- The decision variables in the optimization problem correspond to the design and operational variables in the intermediate-fidelity model (see Section 11.2). In particular, optimal values for both sets of variables are determined simultaneously.

- Important metrics for the comparison of different design and operational alternatives are those expressed in terms of the energy density of the system (see Chapter 8; compare [21]). Typically, the *fuel energy density*, defined as the electrical energy produced per unit mass of fuel, is considered as the performance metric for optimal design and operation. Note that neither the weights of the power generation device nor of the fuel containers are taken into account in this metric, which is therefore best suited for systems where the mass of the fuel dominates (e.g. for long mission durations). Of course, the proposed design and operation methodology is flexible enough that it allows optimization of other metrics [21], such as the overall system energy density for a given mission duration, or the energy efficiency – defined as the power output divided by the product of heating values and molar flow rates.

- Implicit to the optimization formulation is the ability of the intermediate-fidelity model to predict (i) the electrical power produced by the fuel cell, and (ii) the total heat load on the device, for given values of the design and operational variables. Besides meeting the nominal power demand and operating autothermally, the device is typically required to satisfy a number of safety and/ or operational constraints. Safety constraints can be defined to limit the emissions of trace components, such as ammonia or nitric oxide, that have potential impact either on human health or on the environment. For example, these limits can be defined by following the U.S. Occupational Safety and Health Administration (OSHA) regulations[1] or the American Conference of Governmental Industrial Hygienists (ACGIH) recommendations.[2] For structural stability reasons (see Chapter 9 and [22]), the maximum allowable operating temperature may be constrained. Operational constraints may also include limits on the gaseous flow rates in the microchannels, as well as restrictions on the cell potential, depending on the application.

Once an optimal design and operation strategy, that maximizes the device performance and satisfies the constraints, has been determined, a rather natural question that arises is how much confidence can be placed in that strategy? Many sources of uncertainty can indeed invalidate an optimal strategy. Typically, the mathematical expressions used to calculate the reaction rates in an intermediate-fidelity model carry a lot of uncertainty. Moreover, because the geometry of the units is usually not known precisely, certain model parameters may be highly uncertain, such as the catalyst support surface-area-to-volume ratios or the overall heat transfer and emissivity coefficients. Post-optimal sensitivity analysis is a systematic way of analyzing the influence of uncertainty on an optimal strategy [23, 24]. The idea therein is to calculate the variation in the system performance and the optimal design and operation variables incurred by (infinitesimally small) variations in the uncertain parameters. Also, when large parameter variations are considered, parametric programming techniques can be used to monitor the change in optimal solution over the entire parameter range [25].

Besides analyzing the effect of uncertain parameters, parametric studies also prove useful for studying the influence of various design choices or operating

---

1) http://www.osha.gov/SLTC/.
2) http://www.acgih.org/TLV/.

conditions [6]. For example, by varying the nominal power demand and monitoring the corresponding change in performance, one can identify the applications for which the considered technology will most likely be best suited. Regarding design choices, parametric studies can help answer outstanding questions such as: Which reaction kinetics are to be improved in priority for enhancing the system performance? Does the electrolyte thickness have a large influence on the system performance? Is heat recovery between inlet and outlet streams an option worth considering?

The aforementioned considerations are illustrated by a case study subsequently, which is the continuation of the case study presented in Section 11.2.

11.3.1
Case Study (Continued)

Consider the micro power generation process shown in Figure 11.1, for which an intermediate-fidelity model has already been described. The optimal design and operation problem is to determine values of the design variables V^{r}, V^{fc}, V^{bl} and V^{bll} and operational variables T, U, $F_{\mathrm{in}}^{\mathrm{r}}$, $F_{\mathrm{in}}^{\mathrm{ca}}$, $F_{\mathrm{in}}^{\mathrm{bl}}$, $F_{\mathrm{in}}^{\mathrm{bll}}$ that

maximize: fuel energy density
subject to: nominal power demand
autothermal operation
maximum NH_3 and NO emissions.

The objective and constraint functions in this optimization problem are as follows:

- The fuel energy density, $e_{\mathrm{dens}}^{\mathrm{fuel}}$, is calculated as:

$$e_{\mathrm{dens}}^{\mathrm{fuel}} = \frac{\mathcal{P}^{\mathrm{nom}}}{\mathrm{MW}_{\mathrm{NH}_3} F_{\mathrm{in}}^{\mathrm{r}} + \mathrm{MW}_{\mathrm{C}_4\mathrm{H}_{10}} F_{\mathrm{in}}^{\mathrm{bll}}} \tag{11.4}$$

where $\mathcal{P}^{\mathrm{nom}}$ denotes the nominal power demand; $\mathrm{MW}_{\mathrm{NH}_3}$ and $\mathrm{MW}_{\mathrm{C}_4\mathrm{H}_{10}}$ are the molecular weights of ammonia and butane, respectively. Observe, in particular, that maximizing the fuel energy density is equivalent to minimizing the inlet mass flow of ammonia and butane in the system (approximately $F_{\mathrm{in}}^{\mathrm{r}} + 3.4 F_{\mathrm{in}}^{\mathrm{bll}}$).

- The nominal demand constraint simply reads:

$$\mathcal{P} - \mathcal{P}^{\mathrm{nom}} = 0 \tag{11.5}$$

with the electrical power $\mathcal{P}$ produced by the SOFC calculated with Equation 11.3.

- The autothermal operation constraint is given by:

$$\left(\dot{H}_{\mathrm{in}}^{\mathrm{r}} + \dot{H}_{\mathrm{in}}^{\mathrm{ca}} + \dot{H}_{\mathrm{in}}^{\mathrm{bl}} - \dot{H}_{\mathrm{out}}^{\mathrm{bl}} \right) + \left(\dot{H}_{\mathrm{in}}^{\mathrm{bll}} - \dot{H}_{\mathrm{out}}^{\mathrm{bll}} \right) - \dot{Q}^{\mathrm{loss}} - \mathcal{P} = 0 \tag{11.6}$$

Here, $\dot{H}_{in}^{r}$, $\dot{H}_{in}^{ca}$, $\dot{H}_{in}^{bl}$, $\dot{H}_{out}^{bl}$ stand for the inlet and outlet enthalpy flows along the ammonia line, and $\dot{H}_{in}^{bII}$, $\dot{H}_{out}^{bII}$, for the inlet and outlet enthalpy flows along the butane line. They are calculated from the molar enthalpies for pure components and the gas composition and velocities determined by the intermediate-fidelity model. It is assumed that heat recovery between the inlet and outlet gas stream is possible. For example, in the case of half heat recovery, the outlet stream temperature, T^{out}, is equal to $\frac{1}{2}(T - T^{amb})$, with $T^{amb} = 298\,K$ the ambient temperature. Although difficult, note that heat recovery has already been demonstrated in microchemical systems [13].

$\dot{Q}^{loss}$ stands for the overall heat losses to the environment, which takes account of conductive/convective as well as radiative heat losses. In the calculation of $\dot{Q}^{loss}$, a fixed aspect ratio, A^{dev}, is assumed for the device, for example $A^{dev} = 6(V^{r} + V^{fc} + V^{bI} + V^{bII})^{2/3}$ in the case of a cubic box.

- The emission constraints on NH_3 and NO read:

$$y_{NH_3,out}^{bI} \leq y_{NH_3}^{max} \tag{11.7}$$

$$y_{NO,out}^{bI} \leq y_{NO}^{max} \tag{11.8}$$

where the residual molar fractions $y_{NH_3,out}^{bI}$ and $y_{NO,out}^{bI}$ are calculated via the intermediate-fidelity model. The OSHA imposes threshold values of $y_{NH_3}^{max} = 50\,ppm$ and $y_{NO}^{max} = 25\,ppm$, respectively, for these species. However, the ACGIH recommends a tighter time weighted-average (TWA) value of $y_{NH_3}^{max} = 25\,ppm$ in regard to ammonia emissions, which is used instead. It should be noted that the aforementioned threshold limit values are very conservative since they correspond to exposure levels in a typical work environment.

Optimal design and operation results for this process are described below.

11.3.1.1 Optimal Design and Operation for a Nominal Power Demand

For $\mathcal{P}^{nom} = 1\,W$ and in the absence of constraints on the operating temperature, it is found the fuel energy density is maximized for $T = 1445\,K$. Operating at such a high temperature is, however, unrealistic from a practical point of view, mainly because of material stability considerations [22]. Therefore, the operating temperature is removed from the list of decision variables, and a parametric study is carried out by varying its value in a more appropriate range, $T \in [1000\,K, 1300\,K]$.

The Figure 11.3a shows the maximal fuel energy density that is achievable by the process, at steady-state, as a function of the operating temperature. Observe that the maximum achievable energy density is very sensitive to the operating temperature as it varies from $840\,W\,h\,kg^{-1}$ at $1000\,K$, up to $1150\,W\,h\,kg^{-1}$ at $1300\,K$. For the sake of comparison, recall that state-of-the art primary batteries reach up to $700\,W\,h\,kg^{-1}$ and rechargeable batteries up to $300\,W\,h\,kg^{-1}$ [26]. It is also found that, for all operating temperatures in the range 1000–1300 K, the threshold value

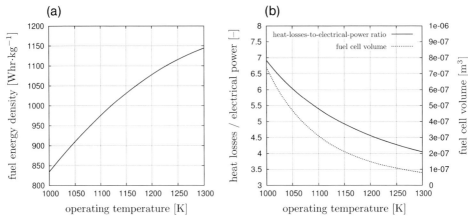

Figure 11.3 Maximum fuel energy density (a), optimal heat-losses-to-electrical-power ratio (b, upper curve), and optimal fuel cell volume (b, lower curve) vs. operating temperature, for $P^{nom} = 1\,W$.

of 25 ppm for NO is attained, whereas the residual concentration of ammonia is negligible.

The effect of temperature on the performance of the system results from a trade-off between the heat losses and the chemical/electrochemical kinetics. On one hand, as temperature increases, so do the chemical/electrochemical reaction rates. In other words, increasing the operating temperature allows one to obtain the same conversions while significantly reducing the size of the units. This behavior is illustrated by the lower dotted curve in Figure 11.3b, where only the optimal fuel cell volume has been represented since it is much larger than the volumes of the other three units at all temperatures. On the other hand, the heat losses per unit area are substantially increased by operating the system at a higher temperature. In this case study, it is found that an increase in the operating temperature actually decreases the heat losses of an optimized device in the range 1000–1300 K, as illustrated by the upper curve in Figure 11.3b. In other words, the additional heat losses per unit area incurred by a higher operating temperature are compensated by a reduction in the optimized device size. This rather counterintuitive result explains why a higher operating temperature is beneficial to the fuel energy density achievable by the system. Note also the very large values of the heat losses to electrical power ratio (the device produces 7 W of heat losses for 1 W of electrical power at 1000 K!) which indicates that a large part of the fuel mass is used to maintain the stack at the prescribed operating temperature.

Regarding operational variables, the optimal value of the air flow rate F_{in}^{bI} in burner I is found to be zero, irrespective of the stack temperature; in other words, it is more efficient to provide all the oxygen needed for the oxidation reactions in burner I through the fuel cell cathode. It is also found that the optimal flow rates of ammonia, air and butane are relatively unaffected by the operating temperature in comparison to the unit volumes; these values decrease by 22, 35 and 37%, respectively, between 1000 and 1300 K – which should be compared to the fourfold decrease in the device volume. The higher consumption of ammonia and butane fuels at lower operating temperatures is directly linked to the decrease in the fuel energy density.

11.3.1.2 Effect of the Nominal Power Demand

The results presented previously are all relative to an electrical power production of 1 W. It is, however, legitimate to raise the question of how the optimal design and operation of the system scales when a power production as low as 0.1 W or, conversely, as high as 10 W is considered. This paragraph provides an overview of that important aspect of the process; compare also Chapter 8.

The effect of the nominal power demand $\mathcal{P}^{nom}$ on the fuel energy density is shown in Figure 11.4a, within the range [0.1–10 W]. The three curves in this plot correspond to operating temperatures of 1100, 1200 and 1300 K. One first sees that the nominal power specification has a very large effect on the performance of the system. At $T = 1300$ K, for instance, the fuel energy density varies from 700 W h kg^{-1} ($\mathcal{P}^{nom} = 0.1$ W) up to 1600 W h kg^{-1} ($\mathcal{P}^{nom} = 10$ W). It is also seen from this figure that operating the system at 1300 K always yields the highest energy densities, irrespective of the nominal power demand.

The relative increase in the heat losses at lower nominal power demands (Figure 11.4b) is directly related to the size of the device itself. In fact, the optimal values of the unit volumes V^r, V^{fc}, V^{bl}, and V^{bll} scale nearly linearly with $\mathcal{P}^{nom}$, whereas the heat losses vary proportionally to $(V^r + V^{fc} + V^{bl} + V^{bll})^{2/3}$. Overall, the heat losses to electrical power ratio of the device thus varies proportionally to $(V^r + V^{fc} + V^{bl} + V^{bll})^{-1/3}$.

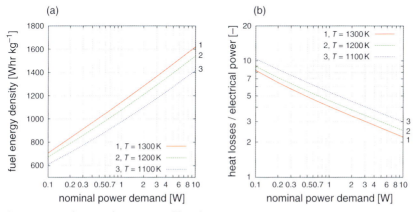

Figure 11.4 Fuel energy density (a) and heat losses to electrical power ratio (b) vs. nominal power demand, for various operating temperatures.

Regarding operational variables, the flow rates of ammonia (in the reactor) and air (in the fuel cell cathode) are found to be nearly proportional to the power demand. It is also found that the increase in the butane flow rate (vs. the power demand) is slower than that of the ammonia flow rate; this is because relatively more butane is needed at lower power demands to compensate for the additional heat losses.

According to the above considerations, it appears that man-portable power generation systems based on high-temperature fuel cells, such as the one shown in Figure 11.1, will most likely be best suited for applications above 1 W. Below this limit, the relative heat losses indeed become so important that these systems may no longer be competitive with existing batteries. These results corroborate those obtained from system-level analysis (see Chapter 8).

11.3.1.3 Sensitivity to Uncertain Kinetic Rates and Resource Allocation

The mathematical expressions used to estimate the kinetic rates in the different unit operations are among the principal sources of uncertainty carried by an intermediate-fidelity model. The following case study considers the influence of uncertain kinetic rates on the optimal design and performance of the fuel cell stack. Here, uncertainty in the kinetic rates is simply represented by varying the pre-exponential factors in the kinetic expressions relative to the chemical/electrochemical reactions in the fuel-processing reactor, the anode and cathode compartments of the SOFC, and the two burners. The nominal power demand is set to $\mathcal{P}^{\mathrm{nom}} = 10\,\mathrm{W}$ and the operating temperature to $T = 1100\,\mathrm{K}$.

Post-optimal sensitivity analysis is used to quantify the effect of this uncertainty on the optimal values of the unit volumes. The results are presented in Figure 11.5 in the form of histograms. For example, in the histogram labeled (a), each bar represents the variation of a given unit's optimal volume in response to a variation in the kinetic rate, r, of NH_3 decomposition $\left(\text{e.g.} \ \dfrac{r^r \partial V^r}{V^r \partial r^r} \right)$. Note that the sensitivity coefficients for the unit volumes are typically negative, since an increase in kinetic rate usually allows one to reduce the volume needed to achieve a given conversion; positive sensitivity coefficients are, however, possible, such as the influence of NH_3 decomposition on V^{bI} since a larger production rate of H_2 leads to a larger H_2 residual concentration at the anode outlet, which in turn justifies a larger burner to generate more heat from the catalytic oxidation of H_2. It is seen that a variation in the kinetic rates of NH_3 decomposition, H_2 and NH_3 oxidation, or C_4H_{10} oxidation, has a very large effect on the optimal volume of the reactor, burner I, or burner II, respectively, but its influence on the volume of other units is very low. On the other hand, a variation in the kinetic rates of electrochemical reaction, either at the anode or cathode of the SOFC, has a large influence, not only on the optimal volume of the SOFC, but also on the optimal volumes of the reactor, burner I, and burner II. Accordingly, more emphasis should be placed on obtaining accurate chemical kinetics for the micro-SOFC than for the other unit operations.

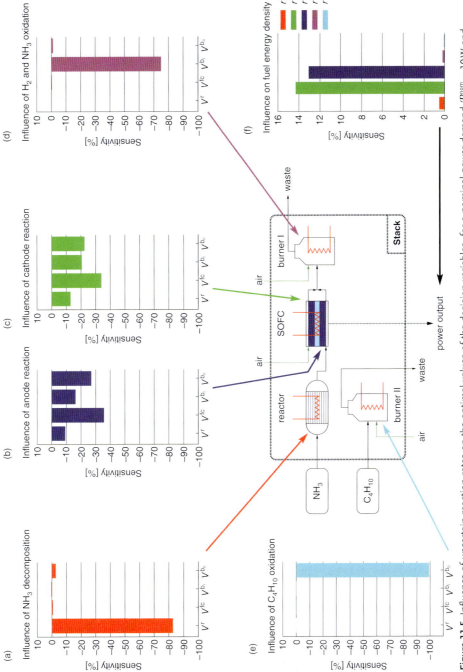

Figure 11.5 Influence of uncertain reaction rates on the optimal values of the decision variables, for a nominal power demand $P^{nom} = 10\,W$ and operating temperature $T = 1100\,K$.

Post-optimal sensitivity analysis is also useful to monitor the influence of the kinetic rates in terms of the system performance. These results are reported in the histogram labeled (f) in Figure 11.5. For example, the green bar represents the variation in energy density incurred by a variation in the kinetic rate of electro-chemical reaction at the cathode side of the SOFC. The largest sensitivity coefficients, of the order of 13–14%, correspond to the kinetic rates at both the anode and cathode of the SOFC. The other sensitivity coefficients are found to be very small, with values less than 1%. Consequently, if one wants to obtain a reliable estimate of the optimal fuel energy density that can be achieved by the process, special care should clearly be taken to develop accurate kinetic models for both half-cell reactions in the SOFC.

The foregoing sensitivity results also provide useful insight to guide the resource allocation problem for component optimization. Suppose, for example, that we are asked to identify which catalyst should be improved in priority to enhance the system performance. Note, first, that the main reason for improved performance is that faster kinetics result in smaller residence time requirements and, therefore, lower heat losses, because of the decreased device size. Because sensitivity analysis has shown that the process performance is mostly sensitive to the fuel cell kinetics, the priority should obviously be given to improving both electrodes simultaneously.

11.4
Optimal Design and Steady-State Operation for Variable Power Demand

Most portable electronic devices are not operated at a constant power demand. Not only the magnitudes, but also the durations of operation at different levels vary. For example, a cellular phone on stand-by mode expends only a small fraction of the power needed during a conversation; the power demand of a portable computer can range between 5 and 30 W depending on usage; the electronic equipment of the dismounted soldier is expected to require 20 W of power with a peak demand of 50 W.

In this section, the optimal design and steady-state operation of devices having a variable power demand is addressed. For simplicity, it is assumed that the power demand can be approximated well through a small number of *discrete* power demands and associated demand frequencies; such an approximation can be made in many practical applications, for example in the list of applications reported previously.

In a variable power demand problem, the demand, in addition to stringent requirements on the operation of the device stemming from reliability and safety considerations, needs to be satisfied for all foreseeable power demand scenarios. In particular, a design based on a single nominal power demand (e.g. as determined previously in Section 11.3) cannot guarantee that such requirements are met for multiple power levels. A conceptually simple way to account for this power variation is to consider the time profile and formulate a dynamic optimization problem based on a transient model. However, this approach is computationally expensive. A more elegant and computationally more tractable approach based on *two-stage stochastic programming* is used subsequently [7]. This formulation has the advantage that only the average time spent at each power demand is needed, in contrast to a detailed transient power profile. A limitation, however, is that the penalty incurred by switching from one power output level to another must be negligible and the duration of transient operation compared to steady-state operation should also be negligible. It is well known that fuel-cell-based power generation systems are not responsive enough to meet rapid demand changes and, consequently, batteries and/or super-capacitors need to be deployed to work in tandem to meet this deficiency (see Chapter 12). In such hybrid electrochemical systems, microfabricated fuel cells are, therefore, intended to operate mostly at steady state, which justifies the foregoing assumptions.

In a two-stage programming formulation, a distinction is made between the design decision variables and the operational decision variables. Essentially, the design decision variables (also called first-stage variables) represent properties of the system that cannot be altered after fabrication of the device (e.g. volume of reactor or fuel cell), while the operational decision variables (also called second-stage variables) represent properties of the system that can be adapted to the current power demand (e.g. operating temperature, cell voltage or flow rates).

- For a given power demand scenario and a specified process design, optimal values for the operational variables can be determined by solving an optimization problem of the form:

  "Determine the operational decisions that maximize the system performance subject to safety and operational constraints."

Note that the latter problem is similar to those solved in Section 11.3 for a nominal power demand, as if the design variables were fixed.

- Optimal values for the design variables, on the other hand, can be determined from the solution of a higher-level optimization problem of the form:

  "Determine the design decisions that maximize the system performance over the set of all possible power demand scenarios, and such that feasible operational decisions can be found for each power demand scenario."

Typically, the performance to be maximized in this problem consists of two contributions: (i) a *design cost* dependent only on the design variables; (ii) an *operating cost* obtained by averaging the operating costs for all foreseeable power demand scenarios. Note that, for each scenario, a set of optimal operational variables is to be determined. In particular, all these operational strategies must satisfy the safety and operational constraints (as described earlier in Section 11.3). Therefore, solving the design and steady-state operation problem in the context of a variable power demand is much more involved than in the nominal case.

Once an optimal design and operation strategy has been identified, parametric studies based either on post-optimal sensitivity analysis (local analysis) or parametric programming (global analysis) can be applied to assess the level of confidence that can be placed in that strategy.

Optimal design and steady-state operation for variable power demand is illustrated by a case study below.

11.4.1
Case Study (Continued)

The man-portable power generation process shown in Figure 11.1 is considered, along with the intermediate-fidelity model described earlier in Section 11.2. It is supposed here that a finite number of possible power demands, $\mathcal{P}_i^{\text{nom}}$, $i = 1, \ldots,$ N_d, can occur during the time frame of interest, with corresponding frequencies of occurrence, ω_i^{nom}. The two-stage programming formulation for optimal design and steady-state operation is to determine the values of the design variables V^r, V^{fc}, V^{bI}, V^{bII}, and, for each possible power demand $\mathcal{P}_i^{\text{nom}}|_{i=1}, \ldots, N_d$, the values of the operational variables T_i, U_i, $F_{\text{in},i}^{r}$, $F_{\text{in},i}^{\text{ca}}$, $F_{\text{in},i}^{\text{bI}}$, $F_{\text{in},i}^{\text{bII}}$ that

maximize: mission fuel energy density
subject to: nominal power demand
autothermal operation
maximum NH_3 and NO emissions $\left.\begin{array}{}\\\\\end{array}\right\}$ for each scenario $i = 1, \ldots, N_d$

A number of remarks are in order:

- The mathematical expressions for the constraints are the same as those given by Equations 11.5–11.8.

- The mission fuel energy density, $e_{\text{dens}}^{\text{mission}}$, is defined as the ratio of total energy produced to the amount of fuel required, normalized by the mission duration. Letting $\mathcal{M}_i$ denote the mass flow rate of fuels in the device corresponding to the power demand $\mathcal{P}_i^{\text{nom}}$, one has:

$$e_{\text{dens}}^{\text{mission}} := \sum_{i=1}^{N_d} \omega_i^{\text{nom}} \mathcal{P}_i^{\text{nom}} \left/ \sum_{i=1}^{N_d} \omega_i^{\text{nom}} \mathcal{M}_i \right. \tag{11.9}$$

This objective is appropriate for applications where the mass of the fuel to be carried dominates over the total mass of the device, or where refueling is infrequent due to logistic restrictions. The numerator of Equation 11.9, being constant for given frequencies and values of the power demand, maximizing $e_{\text{dens}}^{\text{mission}}$ is equivalent to minimizing the sum of mass flow rates weighted by the frequencies.

In this case study, the design of a system to power the electronic equipment of the dismounted soldier is considered. The power demand can take two possible values: 20 and 50 W, with the former occurring 90% of the time [27]. Moreover, the maximum operating temperature for the system is set to 1100 K. The performance of a design for nominal power demand is first considered, then the results obtained with the two-stage programming approach are presented and compared to other design approaches.

11.4.1.1 Performance of Nominal Power Demand Design

In this section, the system is designed based on nominal power demand considerations: first, a nominal power demand is selected and the system is designed for that particular demand, disregarding operation at any other demand; then this design is fixed and optimal operational decisions are determined for the various power demand scenarios. The following approaches are investigated:

- **Design for Average Power Demand**

  The system is designed for the power demand $\mathcal{P}^{\text{ave}} := \dfrac{1}{N_{\text{d}}} \sum_{i=1}^{N_{\text{d}}} \mathcal{P}_i^{\text{nom}} = 35 \text{ W}.$

- **Design for Maximum Power Demand**

  The system is designed for the power demand $\mathcal{P}^{\text{max}} = 50 \text{ W}.$

For performance assessment, the resulting designs are compared to the *ideal design*, which represents a design where the component volumes can be adjusted optimally for each possible power demand. Clearly, this ideal design has the best performance, but it is not realizable.

The results are shown in Figure 11.6. Note that the performances of the designs for average and maximum power vary significantly over the power demand range of interest. As expected, the curves for these designs intersect the ideal design curve at 35 and 50 W demands, respectively. In either case, however, performance

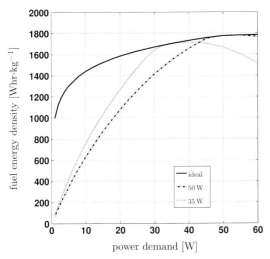

Figure 11.6 Performance of fuel cell stack designs for average power demand (35 W) and maximum power demand (50 W), for the dismounted soldier application.

degrades to unacceptable levels when the actual demand and the design demand differ significantly.

If the actual demand is less than the design demand, the total volume of the device is larger than the volume of the optimal design for that demand. More heat losses occur than is ideal during operation to meet the actual demand and a larger amount of butane than ideally required is thus needed to keep the stack at a constant operating temperature. Accordingly, a decrease in the performance is observed.

If the actual demand is higher than the design demand, the volumes are smaller than the optimal volumes for that demand. Smaller volumes lead to lower conversions in the reactor and fuel cell. In turn, lower conversions are compensated by a higher flow rate of ammonia to provide the necessary H_2 to meet the power demand. The mole fractions of NH_3 and NO in the effluent increase as higher flow rates and smaller volumes result in lower residence times in the units. Larger air flow rates are then required to dilute the mole fraction of NH_3 and NO to acceptable and safe levels. Even though heat losses decrease, more C_4H_{10} is therefore required in order to heat up this additional air. Overall, the flow rates of ammonia and butane increase, and a decrease in the performance is obtained. Even more problematic, as the efficiency of the butane burner decreases with increased C_4H_{10} flow rate (due to shorter residence times), a demand is reached eventually for which the emission constraints on NH_3 and NO can no longer be met by increasing the air flow rates while operating autothermally. For this reason, it not feasible to produce, for example 50 W of electrical power with a 20 W nominal power demand design.

11.4.1.2 Performance of Two-Stage Programming Design
The significant degradation in performance of a nominal power demand design when the actual demand is far from the nominal one motivates the application of the two-stage programming approach to determine design and operational decision variables.

The comparison of the two-stage programming approach with the foregoing nominal power demand designs, in terms of the mission fuel energy density (11.9), is presented in Figure 11.7. Also shown on this figure is the performance of the so-called *conservative design for mean power demand*, which aims to optimize the mean power demand during the mission, $\sum_{i=1}^{N_d} \omega_i^{nom} P_i^{nom} = 23$ W, while guaranteeing that the selected design is feasible at all the power demand levels.

The difference between the ideal design and the two-stage design is only 100 W h kg^{-1}. In particular, the two-stage design allows recovery of a significant portion of the best theoretical possible performance. It is found that this design results in 25% lower fuel expenditure than the maximum power demand design, and 13% lower fuel expenditure than either the conservative mean power demand or the average power demand design. More precisely, it appears that the reduction in fuel expenditure due to the two-stage design is significantly better at lower

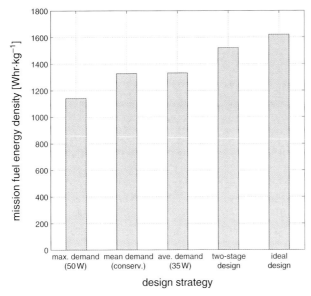

Figure 11.7 Comparison of various design strategies for the dismounted soldier application.

power demands than it is at higher power demands. This is a direct result of the fuel energy density varying more rapidly with demand at lower power demands.

Acknowledgment

This work was supported by the DoD Multidisciplinary University Research Initiative (MURI) program administered by the Army Research Office under Grant DAAD19-01-1-0566. The author would like to acknowledge P. I. Barton, A. Mitsos, and M. Yunt for their contributions which led to this chapter.

References

1 Mitsos, A., Hencke, M.M. and Barton, P.I. (2005) Product engineering for manportable power generation based on fuel cells. *AIChE Journal*, **51** (8), 2199–219.

2 Mitsos, A., Palou-Rivera, I. and Barton, P.I. (2004) Alternatives for micropower generation processes. *Industrial and Engineering Chemistry Research*, **43** (1), 74–84.

3 Deshmukh, S.R. and Vlachos, D.G. (2005) Effect of flow configuration on the operation of coupled combustor/reformer microdevices for hydrogen production. *Chemical Engineering Science*, **60**, 5718–28.

4 Hsing, I.M., Srinivasan, R., Harold, M.P., Jensen, K.F. and Schmidt, M.A. (2000) Simulation of micromachined chemical reactors for heterogeneous partial oxidation reactions. *Chemical Engineering Science*, **55**, 3–13.

5 Mitsos, A., Chachuat, B. and Barton, P.I. (2007a) Methodology for the design of man-portable power generation devices.

Industrial and Engineering Chemistry Research, **46** (22), 7164–76.

6 Chachuat, B., Mitsos, A. and Barton, P.I. (2005a) Optimal design and steadystate operation of micro power generation employing fuel cells. *Chemical Engineering Science*, **60** (16), 4535–56.

7 Yunt, M., Chachuat, B., Mitsos, A. and Barton, P.I. (2008) Designing manportable power generation systems for varying power demand. *AIChE Journal*, **54** (5), 1254–69.

8 Barton, P.I., Mitsos, A. and Chachuat, B. (2005) Optimal start-up of micro power generation processes, in *Computer Aided Chemical Engineering*, Vol. 20B (eds L. Puigjaner and A. Espuña), Elsevier, The Netherlands, pp. 1093–8.

9 Chachuat, B., Mitsos, A. and Barton, P.I. (2005) Optimal start-up of micro power generation processes employing fuel cells, in *AIChE Annual Meeting, 30 Oct–4 Nov 2005, Cincinnati, OH.*

10 Norton, D.G. and Vlachos, D.G. (2004) A CFD study of propane/air microflame stability. *Combustion and Flame*, **138**, 97–107.

11 Mitsos, A. (2006) Man-Portable Power Generation Devices: Product Design and Supporting Algorithms. PhD thesis, Massachusetts Institute of Technology, Cambridge, MA.

12 Poshusta, J.C., Kulprathipanja, A., Martin, J.L. and Martin, C.M. (2006) Design and integration of portable SOFC generators, in *AIChE Annual Meeting, San Francisco, CA, 12–17 Nov 2006.*

13 Arana, L.R., Baertsch, C.D., Franz, A.J., Schmidt, M.A. and Jensen, K.F. (2003) A microfabricated suspended-tube chemical reactor for thermally efficient fuel processing. *IEEE Journal of MEMS*, **12** (5), 600–12.

14 Fogler, H.S. (1998) *Elements of Chemical Reaction Engineering*, 3rd edn, Prentice-Hall, New Jersey.

15 Deshmukh, S.R., Mhadeshwar, A.B. and Vlachos, D.G. (2004) Microreactor modeling for hydrogen production from ammonia decomposition on ruthenium. *Industrial and Engineering Chemistry Research*, **43** (12), 2986–99.

16 Bessette, N.F., II, Wepfer, W.J. and Winnick, J. (1995) A mathematical model of a solid oxide fuel cell. *Journal of The Electrochemical Society*, **142** (11), 3792–800.

17 Achenbach, E. (1994) Three-dimensional and time-dependent simulation of a planar solid oxide fuel cell. *Journal of Power Sources*, **49**, 333–48.

18 Pignet, T. and Schmidt, L.D. (1975) Kinetics of NH_3 oxidation on Pt, Rh and Pd. *Journal of Catalysis*, **40**, 212–25.

19 Westbrook, C.K. and Dryer, F.L. (1984) Chemical kinetic modeling of hydrocarbon combustion. *Proceedings of the Combustion Institute*, **10**, 1–57.

20 Biegler, L.T. and Grossmann, I.E. (2004) Retrospective on optimization. *Computers & Chemical Engineering*, **28** (8), 1169–92.

21 Mitsos, A., Chachuat, B. and Barton, P.I. (2007b) What is the design objective for portable power generation: Efficiency or energy density? *Journal of Power Sources*, **164** (2), 678–87.

22 Srikar, V.T., Turner, K.T., Ie, T.Y.A. and Spearing, S.M. (2004) Structural design considerations for micromachined solid-oxide fuel cells. *Journal of Power Sources*, **125** (1), 62–9.

23 Buskens, C. and Maurer, H. (2000) SQP-methods for solving optimal control problems with control and state constraints: adjoint variables, sensitivity analysis and real-time control. *Journal of Computational and Applied Mathematics*, **120** (1–2), 85–108.

24 Fiacco, A.V. (1983) *Introduction to Sensitivity and Stability Analysis in Nonlinear Programming, volume 165 of Mathematics in Science and Engineering*, Academic Press, New York.

25 Bank, B., Guddat, J., Klatte, D., Kummer, B. and Tammer, K. (1983) *Non-Linear Parametric Optimization*, Birkhäuser Verlag, Basel.

26 Linden, D. (2001) *Handbook of Batteries*, McGraw Hill.

27 Soldier Power/Energy Systems Committee (2004) *Meeting the Energy Needs of Future Warriors*, The National Academies Press, Washington DC.

12
Design of Hybrid Electrochemical Devices

Andrew T. Stamps and Edward P. Gatzke

12.1

Introduction

This chapter considers the design of hybrid electrochemical devices. Hybrid electrochemical devices are composed of multiple different power sources. In many applications, a single type of power supply may not be adequate due to performance characteristic limitations. These limitations are typically related to maximum deliverable power, the total energy available from the power supply, or other physical characteristics such as weight, volume, and cost. As a result, a hybrid system composed of two or more disparate power sources can be used to meet design goals which cannot be achieved using a single power supply. Note that this chapter does not consider switched hybrid dynamic systems [1] or hybrid control systems [2], although these types of approaches may also be relevant in part to hybrid electrochemical systems.

Consider two electrochemical power supply components: a fuel cell and a capacitor. A fuel cell may have a high energy density, but fuel cell systems often have limited power density. In practical terms, this means that fuel cells can provide low levels of energy for extended periods of time. In contrast, electrochemical capacitors can supply high power but contain limited energy. Hybrid electrochemical power systems mix multiple components, allowing operation under conditions not attainable by single component systems. In some operating conditions, hybrid electrochemical power systems can operate with power and energy density not achievable by single components alone.

The benefits of hybrid power systems are truly realized for dynamic systems with time-varying loads. This complicates the design process, as the designer must have some description of the expected load profile. Typically, this is provided as a base load level with peak usage and periodic cycle times specified, although variable power demands with assumed random characteristic could also be considered in the chapter devoted to optimal design and operation. The example from this chapter considers a simple repeating load cycle with two load levels, but this profile could be readily extended to a more complex profile with numerous load levels.

Microfabricated Power Generation Devices. Edited by Alexander Mitsos and Paul I. Barton
Copyright © 2009 WILEY-VCH Verlag GmbH & Co. KGaA, Weinheim
ISBN: 978-3-527-32081-3

Most electrochemical components such as batteries and capacitors are inherently dynamic systems. These components do not have a nominal steady-state operating voltage or power rating. In contrast, fuel cell systems typically operate at or near steady state, converting the chemical potential in liquid or gaseous fuels to electrical power. Batteries and capacitors are energy storage devices, not energy production devices. As a result, batteries and capacitors are only at steady state when they are not being used (although some components may actually self-discharge while not in use). Dynamic modeling and optimization of dynamic systems are key techniques that must be considered when designing hybrid electrochemical systems. Therefore, effective design necessitates the development and use of dynamic models for these components.

This chapter is organized as follows: Previous efforts in the area of modeling and design of hybrid electrochemical power systems are first considered. Then, numerical methods for optimal design are examined with a specific consideration for optimization of dynamic systems. An application of numerical methods for design of an example system is discussed. Finally, future considerations for design of hybrid power systems are presented.

12.2
Hybrid Electrochemical Systems

While many researchers have examined hybrid systems in the past from different directions, no single tool or method has been created that is available to design and optimize hybrid power systems in a general manner. A variety of authors have approached different aspects of hybrid system modeling. Some have used hybrid power system models for design.

One of the simplest hybrid power systems is the use of capacitors in parallel with batteries. Burke [3] and Miller [4] examined this type of hybrid system with batteries connected in parallel to capacitors. This was found to improve the power capability of the batteries. Similar studies were carried out on batteries in parallel with capacitors under pulse loads, and these hybrid systems demonstrated increased power capability as well as the extended run time of the battery [5–7]. Additional studies on battery/electrochemical hybrid systems were performed by several other research groups for a wide range of applications [8–10]. Even simple hybrid power systems can improve system performance.

Fuel cell power systems have been the subject of significant research in recent years. The concept of hybridization using fuel cells is an active area of research, as the fuel cell cannot meet high power demands. Hybrid fuel cell systems are used in many applications such as submarines, automobiles, and portable devices. Nadal *et al.* [11] developed a proton exchange membrane (PEM) fuel cell/lead acid battery hybrid system to power a lightweight passenger vehicle. Jarvis *et al.* [12] reported that a PEMFC stack could not supply a power demand of $18\,W$ for more than $2\,s$, while a $70\,F$ capacitor bank in parallel to the PEMFC stack was able to supply the same power demand. Jiang *et al.* [13] designed control strategies for power sharing between a Li-ion battery and a PEM fuel cell. Kim and Peng [14] examined the optimization of a fuel cell/battery hybrid vehicle system using a numerical simulation of the dynamic system model, including development of power sharing control methodologies. Considering these results, research into hybrid systems using electrochemical system components indicates great potential and has grown significantly. However, limited work has appeared presenting general methods for modeling and optimizing portable hybrid power systems.

Many references for hybrid power systems describe only the application with little attention to modeling. In the context of systems modeling, Dougal *et al.* [15] presented analytical solutions to a simplified model of a battery–capacitor hybrid system based on circuit modeling and discussed in detail the energy efficiency, power capabilities and current sharing between the battery and the capacitor. Their group also developed a circuit-based model to predict the performance of a portable battery–fuel cell system. Zheng *et al.* [5] presented a simple theoretical analysis of an ideal capacitor and an ideal battery in parallel and derived a relation between the current shared by the battery and the capacitor. A more sophisticated physics-based model for a battery–electrochemical capacitor was developed by Sikha *et al.* [9] and simple optimization strategies using theoretical energy and power density relationships were discussed. However, none of these previous investigators presented a methodology to find the optimal design of the hybrid system using physics-based component models and a specified power profile with path constraints for components.

Historically, a resistive companion approach has been considered for the general purpose modeling of systems integrating electrical components. In resistive companion component models, the voltage difference for each component is specified and the current for that component is then calculated. This methodology greatly simplifies simulation of electronic devices, as current is easily determined for each component. Although this resistive companion approach works well for systems with two-terminal devices (components with two electrical connections such as batteries and capacitors) it does not readily extend to multi-terminal devices such as transistors, power converters, and integrated circuits. The potential difference for the component may lead to less accurate models due to the numerous unique potential differences between terminals within a single component, which may influence the behavior of the component. Nevertheless, this modeling framework can be forced to accommodate more general components. Namely, potentials are mapped to terminals on the component and then passed to the component model where the appropriate potential differences are computed internally. Furthermore, universal sign conventions are typically adopted for flow of current into terminals. Thus, all currents for terminals connected to a common potential can be summed to zero without having to consider whether a terminal is "positive" or "negative". Various packages including Cadence PSpice [16], The MathWorks SimPowerSystems [17], and the Virtual Test Bed [18] developed out of the University of South Carolina have all incorporated this sort of approach. More general simulation methodologies for modeling are available, but these methods usually require greater domain knowledge from the user since component models for electrochemical systems are not commonly available by default. These general simulation methods usually rely on a differential algebraic equation (DAE) model representation. Many DAE solution methods are available, many of which trace their origin to DASSL/DASPK [19].

12.3
Simultaneous Optimization of Dynamic Systems

When designing a system, one strives to find the best solution to the problem. As stated in previous chapters, the designer must specify some information before attempting to solve the overall design problem. Design variables such as component size and type may be considered, as well as the control methodology. The overall design criteria must be specified by the designer. For specialty portable devices, minimum weight and/or volume may be the primary objective, whereas cost may be the critical objective for a commodity device; often a weighted combination of objectives is desired. Additional limitations on the design are frequently required. These limitations include design specifications such as power level and bus voltage or physical limits on component size/weight. Constraints also arise to ensure the safety of the system or to protect individual components limits such as enforcing minimum/maximum voltages across components or maximum component currents. Component model type must also be considered. In some cases, simple algebraic models may suffice, while other cases call for advanced dynamic models represented by ordinary differential equations (ODEs), DAEs, or partial differential equations (PDEs). Once these initial issues are addressed, a mathematical formulation for a numerical optimization design problem can be considered.

Numerical optimization can be used to determine the best overall design, given component models and design limitations. Numerous optimization formulations are available, but most types fall into the class of constrained optimization problems. There are many equivalent representations of these constrained problems; one such representation is given by Equation 12.1 below:

$$\min_x \varphi(x)$$
$$\text{s.t.} \quad g(x) \leq 0 \tag{12.1}$$
$$h(x) = 0$$
$$x^L \leq x \leq x^U$$

Most current numerical optimization solution methods operate on algebraic objective functions and constraints. Numerical optimization methods are often used in engineering applications such as parameter estimation, model identification, optimal control, and process design. These applications often involve dynamic systems. Considerable effort has been devoted to formulating problems involving dynamic systems so that they can be solved by traditional numerical optimization methods. The primary difficulty in this process is that dynamic equations represent constraints that must be enforced at an infinite number of points within a specified time interval. The optimization problem transforms from a finite-dimensional variable space to an infinite one. The direct solution of a problem with an infinite number of variables is typically not possible. Thus, approximations are made, depending on the situation, to convert the infinite-dimensional problem to a finite one that can be solved using established methods.

There are several classes of dynamic models and a variety of ways to represent them mathematically, but for the purpose of this work it is assumed that the system can be described as a set of linearly-implicit DAEs as:

$$M\frac{dx}{dt} = f(t, x, p) \qquad (12.2)$$

whose dynamic behavior depends on the design variables p. The linearly-implicit DAE form also includes the so-called *mass matrix M* that is assumed to have constant coefficients. When M has full rank, the equations comprise a set of ODEs; a singular matrix denotes a DAE system. The optimal design problem can then can be formulated as the following *Dynamic Optimization* problem:

$$
\begin{aligned}
\min_{x,p} \quad & \varphi(x(t), p) \\
\text{s.t.} \quad & M\frac{dx}{dt} - f(t, x(t), p) = 0 \quad \forall t \in [t_0, t_f] \\
& g(x(t), p) \leq 0 \qquad\qquad \forall t \in [t_0, t_f] \\
& h(x(t_0), p) = 0 \\
& x^L \leq x(t) \leq x^U \qquad \forall t \in [t_0, t_f] \\
& p^L \leq p \leq p^U
\end{aligned}
\qquad (12.3)
$$

The primary challenge in this optimization problem formulation in Equation 12.3 is that the finite number of equality and inequality constraints must be satisfied at an infinite number of points in the continuous interval $[t_0, t_f]$ for the system defined in Equation 12.2. A problem with an infinite number of variables generally cannot be solved directly. Hence, a number of different approaches have been developed to approximate this important class of problems with finite-dimensional representations that can be solved by existing methods. These approaches are categorized in various ways including *variational* methods, *sequential* or *control variable parametrization* methods, *multiple shooting* and *quasi-sequential* methods, and *simultaneous* or *direct transcription* methods.

A simultaneous approach was used for the subsequent example. Unlike a number of the other methods, the simultaneous approach applies a discretization method in the time domain to the state variables. This procedure converts the continuous-time DAE equations into a set of algebraic equations which can be represented as constraints in the optimization formulation. Collocation on finite elements is the preferred discretization as it has a number of desirable stability and convergence properties. The collocation method has been used successfully for engineering applications for quite some time [20] and is most commonly used as a technique for solving two-point boundary value problems. Basic formulations are covered in numerous locations including [21–23]. However, it can be used effectively for initial value problems as well, particularly when the collocation points are chosen to be the Radau quadrature points. A polynomial approximation for each unknown state is constructed on a known number of intervals (finite elements) using a trial function and the derivative of the trial function. The function and its derivative can be evaluated at multiple points within the collocation element. Thus, the DAE governing equations are converted to a set of algebraic equations, albeit very large. The scale of the resulting nonlinear programming (NLP) optimization problem is a major concern, as discretization of complex DAE systems will generally create very large problems. However, modern NLP solution techniques are capable of solving problems involving many thousands of variables and constraints.

By replacing the continuous constraints in Equation 12.3 with the collocation equations, evaluating path constraints at the collocation points, and including the state values $x_{k,j}$ directly in the optimization formulation, the dynamic optimization problem is converted to a standard NLP of the form given by Equation 12.1. Note that use of path constraints with discretized systems can potentially allow for constraint violation between the discretization points. In the example used in this chapter, three-point collocation has been selected, which results in piecewise cubic trial solutions. Ultimately, this procedure produces a numerically stable system of equations and exhibits favorable properties for the solution of stiff ODEs or DAEs [24]. Moreover, with appropriate assumptions, it can be shown that the necessary conditions for optimality of the reformulated problem approach those for the original problem generated using variational methods as the number of collocation points increases [25].

12.4
Hybrid Power System Optimization

Consider a simple hybrid power system consisting of a lithium ion battery, electrochemical capacitor, and PEM fuel cell in parallel with a load, as depicted in Figure 12.1. Load sharing is accomplished passively based on the impedance of the respective components. This system is nearly the simplest hybrid power system one may consider (a passive two-component system being simplest). The system load is specified *a priori* as current or power as a function of time. Simple approaches using average component performance may not be adequate for applications with varied loads. Dynamics from load transients can cause significant changes in component voltage and power levels, especially when using passive power sharing. Active power sharing for DC–DC conversion may help mitigate some problems found in hybrid systems but will increase system complexity, costs, and inefficiency.

Despite the simplicity of the system, five parameters can be manipulated when designing this system. Assuming a battery pack is considered with N_{series} identical

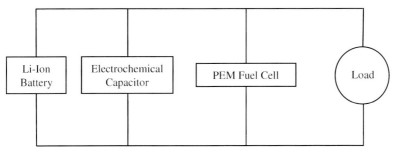

Figure 12.1 Example hybrid power system schematic.

batteries in series and $N_{parallel}$ identical strings of batteries in parallel, the design parameters are: (i) the number of cells in series in the battery N_{series}, (ii) the number of cells in parallel in the battery $N_{parallel}$, (iii) the number of capacitors in parallel N_{cap}, (iv) the number of cells in the PEMFC stack N_{cells} and (v) the membrane area of each fuel cell A_{FC}. Although the first four parameters would be integer values in any hardware implementation, they are treated as continuous variables for the purpose of modeling and optimization within this example. The resulting design with continuous variables can be implemented in two ways. Specialized components could be created to exactly match the required battery or capacitor capacity. Alternatively, the suggested design values could be rounded up to the next integral value, resulting in a sub-optimal design. Included in this example are the models of the individual components as well as the design problem formulation and analysis of the results.

12.4.1
Li-Ion Battery

This model has been used previously for capacity fade analysis [26]. In this model, the state of charge (SOC) of the battery θ is modeled as a function of time given by the following equation:

$$\frac{d\theta}{dt} = -\frac{I_{batt}(t)}{Q_{batt}}$$

where $I_{batt}(t)$ is the current (in amperes) being provided by the battery as a function of time and Q_{batt} is the nominal capacity of the cell (in coulombs). Assumption of a constant Q_{batt} allows design using standard-sized cells with a known voltage range. The operating voltage of the cell at any time is given by the algebraic relationship

$$V_{batt}(t) = U(\theta(t)) - RI_{batt}(t)$$

where R is a lumped resistance which accounts for multiple resistive losses within the cell and $U(\theta)$ is an empirical correlation relating cell potential to the SOC. As was done in [26], $U(\theta)$ is chosen to be a 9th order polynomial

$$U(\theta) = \sum_{j=0}^{9} a_j \theta^j$$

calibrated against low-rate discharge data. Recalling that the battery pack has N_{series} identical batteries in series and $N_{parallel}$ identical strings of batteries in parallel, the model equations are then given as:

$$\frac{d\theta_{pack}}{dt} = -\frac{I_{pack}(t)}{N_{paralled} Q_{batt}}$$

$$V_{pack}(t) = N_{series}\left(U(\theta(t)) - R\frac{I_{pack}(t)}{N_{parallel}} \right)$$

This model is valid for both charging (positive I_{pack}) and discharging (negative I_{pack}) of the battery.

This dynamic battery model is effectively a single nonlinear ODE representation of a battery. Although quite simple, this battery model has been found to accurately represent real battery systems in a variety of experimental conditions. The non-linear relationship between voltage and SOC makes this an interesting dynamic problem, especially when combined with other components in a hybrid system using passive power sharing. Additionally, over voltage constraints could be incorporated in the design formulation to avoid transient operation in dangerous conditions during pulsed power operation.

12.4.2
Capacitor

The electrochemical capacitor will be approximated by an ideal capacitor with a large capacitance in series with a small resistance.[1] The governing equation of a capacitor is given by

$$I_{cap} = C\frac{dV_{cap}}{dt} \tag{12.4}$$

where C is the capacitance of the device and V_{cap} is the voltage drop across the device. Since the system modeling approach requires the specification of the potential difference across a component and not the derivative of the potential difference, Equation 12.4 is integrated and rearranged to provide the appropriate relationship:

$$V_{cap}(t) = V_{cap}(0) + \frac{1}{C}\int_0^t I_{cap}(\tau)d\tau$$

$$= V_{cap}(0) + \frac{1}{C}(q_{cap}(t) - q_{cap}(0)) \tag{12.5}$$

[1] The resistance is both physically relevant and assists in the numerical convergence and stability of the capacitor model.

where q_{cap} is the electric charge stored in the capacitor and τ is a dummy integration variable. The initial voltage $V_{cap}(0)$ is typically treated as a parameter, but specifying $q_{cap}(0) = 0$ and requiring steady-state initialization forces $V_{cap}(0)$ to be zero. After accounting for the internal resistance R_{cap}, Equation 12.5 becomes

$$V_{cap}(t) = V_{cap}(0) - I_{cap}(t)R_{cap} + \frac{1}{C}(q_{cap}(t) - q_{cap}(0)) \tag{12.6}$$

Thus, the capacitor is modeled with the differential equation on charge q_{cap}:

$$\frac{dq_{cap}}{dt} = -I_{cap}(t)$$

and the algebraic constraint given by Equation 12.6.

The capacitor potential, V_{cap}, is specified using Kirchoff's current law in the overall system model representation. For the design problem, the system capacitance is specified by the number of 10 F electrochemical capacitors in parallel, $C = 10N_{cap}$. Moreover, the maximum operating voltage of electrochemical capacitors is such that two sets of capacitors must be connected in series for each battery in series. This means the overall number of capacitors is given by:

$$N_{tot\ cap} = 2N_{cap}N_{series}$$

Thus, based on the general rules for connecting capacitors in series and parallel, the capacitance C (in F) for the model is given by the following equation:

$$C = 5\frac{N_{cap}}{N_{series}}$$

12.4.3
Fuel Cell

The fuel cell model considered in this work is an empirical correlation model developed by Kim *et al.* [27]. This steady-state model provides an algebraic relationship between the current density of the cell i_{FC} and the potential of the cell E_{FC}, as shown in Equation 12.7:

$$E_{FC} = E_0 - b\log i_{FC} - R_{FC}i_{FC} - m\exp(ni_{FC}) \tag{12.7}$$

There are five adjustable parameters in this model: E_0, the equilibrium potential of the cell; b, the Tafel slope associated with activation losses, primarily in the cathode; R_{FC}, the ohmic loss largely related to membrane properties; and m and n, two parameters associated with mass transport limitations at high load conditions. Although the first three parameters have physical meaning and can be estimated from theory, all five parameters are typically used to fit the data of a particular cell at given operating conditions including temperature, pressure, humidity, and cathode composition level (e.g. O_2 vs. air cathode).

This relatively simple five-parameter model has been shown to be sufficient to fit PEMFC polarization curves taken over a very wide range of operating conditions [27]. However, each set of parameters obtained can only be used for I–V prediction under the specific set of operating conditions for which they were calibrated. Nevertheless, this model can be scaled up to approximate the I–V performance of a stack with perfect temperature, humidity, and pressure control through the following equations:

$$i_{FC} = I_{stack}/A_{FC}$$

$$V_{stack} = N_{cells}E_{FC}$$

When used in this work, the reported parameters for fully-humidified 1 atm air cathode operation at 70 °C are used. Note that more complicated fuel cell models based on first principles constitutive relationships could be used, but additional model accuracy is offset by computational complexity for simulation and optimization. Additionally, constraints could be incorporated to limit the operating regime of the fuel cell system to minimize problems associated with high current or voltage.

12.4.4
Programmable Load

In this example, the load is a periodic pulsed power profile, exemplified in Figure 12.2. Note that the pulse is fully characterized by five parameters: the base power level P_{base}, the maximum power level P_{max}, the period length t_{period}, the fraction of time (duty) at peak power d, and the offset time of the peak from the beginning

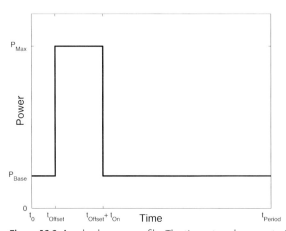

Figure 12.2 A pulsed power profile. The time at peak power t_{on} is given by $d\, t_{period}$.

of the cycle t_{offset}. While the signal is theoretically discontinuous, hyperbolic tangents have been used to generate a continuous approximation for use with the simultaneous dynamic optimization method. This approximation appears to be reasonable and accurate for the given application with limited effect on the results. Given the desired power profile as a function of time $P(t)$, the governing equation for the load is

$$P(t) = V_{sys}(t)I_{sys}(t)$$

where the $V_{sys}(t)$ is the system voltage and the system current $I_{sys}(t)$ is given by $I_{pack}(t) + I_{cap}(t) + I_{stack}(t)$.

12.4.5
Problem Formulation and Results

The target application of this example system is man-portable power. Therefore, the design optimization is formulated to minimize system mass. One factor that heavily influences the design size is the power specifications. In order to examine the sensitivity of the design with respect to these specifications, designs are computed for a variety of pulse characteristics. Specifically, the period, duty, and peak power of the pulse are varied as shown in Table 12.1. The base power is kept constant at 50 W for all trials. Full factorial sampling of the pulsed-power parameter values yields 448 different combinations, resulting in many numerical design problems to be solved.

The form of the system mass objective function is given by Equation 12.8:

$$\varphi_W = W_{Batt}N_{series}N_{parallel} + 2W_{Cap}N_{series}N_{cap} + W_{FC}N_{cells}A_{FC} + W_{Pen} \tag{12.8}$$

where W_{Batt}, W_{Cap}, and W_{FC} are the weighting factors for the battery pack, capacitor bank, and fuel cell stack, respectively. Here, W_{Pen} is a penalty function used to ensure that the batteries and capacitors are recharged at the end of the period. The value of W_{Batt} was estimated to be 43 g for a 1.4 A h Li-ion cell, obtained by weighing a Sony 18650 Li-ion battery. The value W_{Cap} was 0.5 g for a 10 F electrochemical capacitor, a rather small value. Given the relatively small weight contribution of

Table 12.1 Pulsed-power load permutations. Designs were investigated for the 448 different power profiles resulting from all combinations of the three parameters listed below.

Parameter	Values
Period (s)	{10, 15, 20, 25, 30, 45, 60, 120}
Duty	{0.02, 0.05, 0.10, 0.15, 0.20, 0.25, 0.35, 0.50}
Peak power (W)	{100, 125, 150, 200, 250, 350, 500}

the capacitors in the final designs and the relatively constant number across all designs, an increased unit mass would only further reduce the number of capacitors. The incremental weight for the fuel cell stack is approximated as $0.35\,g\,cm^{-2}$ per cell, a value based on estimates from existing commercial fuel cell stacks. Note that the weight contribution of the capacitor bank contains a factor of $2N_{series}$. The standard voltage rating for electrochemical capacitors is about one half that of Li-ion batteries, so it is assumed that the capacitor bank is assembled with two capacitors in series for each battery in series in the battery pack. This not only ensures that the capacitors are not operated outside their voltage rating, but also reduces the capacitance of the overall capacitor bank.

In addition to minimizing weight, the resulting design must satisfy additional constraints. The device is to be designed to supply 12 VDC unregulated power. In other words, the design is acceptable as long as the system voltage remains greater than 12 V. Thus, the following constraint is imposed at all discretization points in the formulation:

$$V_{sys} \geq 12 \text{ V}$$

This lower bound on voltage is an arbitrary design constraint, motivated by a need for the load to maintain this voltage. Upper bounds could also be imposed, but the design problem could become infeasible in cases with incompatible constraint specifications. From a physical standpoint, the types of equipment for which this system is designed are capable of operation over a wide range of voltages, provided the minimum voltage limit is satisfied. Furthermore, the weight-minimization objective is sufficient to ensure that exceedingly high voltages are not delivered.

In addition to the performance specification, the batteries and fuel cells have safe operating limits beyond which these devices fail or their useful operating life is significantly reduced. The voltage across a single Li-ion battery (containing a $LiCoO_2$ cathode and hard carbon anode) should remain in the range 2.8–4.2 V, which results in the following path constraints at each discretization point:

$$2.8 \text{ V} \leq \frac{V_{sys}}{N_{series}} \leq 4.2 \text{ V}$$

Similarly, the batteries should not be charged or discharged at excessive rates. Charging is limited to a 1 C rate while discharge was capped at 3 C, which yields

$$-1.4 \text{ A} \leq \frac{I_{Batt}}{N_{parallel}} \leq 4.2 \text{ A}$$

assuming that charging currents are negative and a 1.4 Ah nominal capacity battery. Likewise, an individual PEM fuel cell should operate in a voltage range of about 0.3–1.1 V, which is equivalent to the following constraints:

$$0.3 \text{ V} \leq \frac{V_{sys}}{N_{cells}} \leq 1.1 \text{ V}$$

Finally, the current density on the fuel cell is also limited to positive values and capped to a maximum value:

$$10^{-3}\,\text{Acm}^2 \le \frac{I_{FC}}{A_{FC}} \le 1.5\,\text{Acm}^2$$

These eight path constraints along with the lower bound constraint on system voltage ensure that computed designs will operate within specified performance constraints without unacceptable stresses on the individual components. Additionally, constraints are imposed to force the design to steady-state initial conditions so that initial equilibration dynamics do not influence design results. An additional objective penalty term was included in order to recharge the batteries and capacitors by the end of the pulsed power period.

After some experimentation, the time interval for each design optimization was divided into 200 equally-spaced elements. The three-component models and the load comprise a system of six DAEs, including two differential and four algebraic states. Applying three-point collocation on each of the 200 elements to the system model and including the initial conditions results in 3606 state variables x in the optimization problem formulation. The eight path constraints applied at each discretization point result in the creation of an additional 4800 slack variables s. Finally, there are the five design parameters p which consist of N_{series}, $N_{parallel}$, N_{cap}, N_{cells}, and A_{FC}. A set of MATLAB scripts developed by one of the authors was used to automate the collocation and problem formulation procedure. There are a total of 8411 equations and 8406 constraints in the formulation. Solution times for the 448 design problems ranged from 150–900 s on a 2.4 GHz Pentium 4 computer with 1.5 GB of RAM using a MATLAB interface to the Fortran version of IPOPT, v 2.3 [28]. The limited-memory BFGS Hessian approximation option was utilized with a 25-step history instead of the analytic Hessian, and the maximum number of second-order corrections was reduced to 1.[2] The remaining tolerances, parameters, and options were left at their default values.

The overall system weights obtained from these designs are shown in Figure 12.3. The different surfaces plotted correspond to the different maximum power values P_{max}. Not surprisingly, as P_{max} is increased the overall design weight increases. However, the cycling period was found to have a negligible effect on design weight for the range of periods studied. While initially puzzling, this is believed to be due to the specification of the pulse time in terms of the duty ratio instead of an absolute length, which results in a constant average power for a given duty, independent of period length, while maintaining a constant ratio of P_{max} to P_{base} as given by Equation 12.9:

$$P_{avg} - \frac{1}{\tau_p}\int_0^{\tau_p} P(t)\mathrm{d}t = \mathrm{d}P_{max} + (1-d)P_{base} \tag{12.9}$$

---

2) Testing by the authors found that utilizing the exact Hessian increased solution times due to the high cost of evaluating their Hessian function, despite reducing the number of iterations needed to achieve convergence.

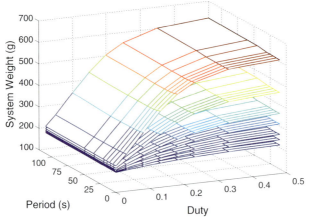

Figure 12.3 System design weight for various pulse-power design specifications. The surfaces correspond to the different values of P_{max} studied where $P_{max} \in$ {100, 125, 150, 200, 250, 350, 500}.

where τ_p is the length of the period and d is the duty. The effect of the period is not identically zero numerically. The period was considered as a parameter because the capacitor may sustain more of the load at higher frequencies/shorter periods. However, the range presented did not reach high enough frequencies to show this trend. It is expected that as the frequency exceeds 1 Hz that a capacitor effect may be present.

Since the period is observed to have minimal effect, the data are replotted for a fixed-period slice considering only variations in maximum power and duty. The data corresponding to the median period value of 30 s are selected, and these results are shown in Figure 12.4. Now it can be seen that the system weight depends nearly linearly on the maximum power, although the slope of this dependence varies significantly with the duty. However, the trend pertaining to the duty appears to be more nonlinear. At small values, the weight increase appears linear, but then begins to approach a plateau asymptotically at higher values of d. This indicates that the influence of P_{max} is more significant to design weight than simply its contribution to the average power. If that were the case, the dependence would be strictly linear. Consequently, average power cannot be used as a successful predictor of design weight scaling by itself. This is illustrated by the sloped characteristics of the *average* power contours in the figure – optimal designs for systems providing the same average power do not necessarily have the same optimal weight! Obviously, attempting to classify a dynamic power profile by a single quantity and neglecting the dynamic characteristics of the load has potentially costly ramifications in the design phase; a design runs the risk of either being overweight or unable to provide the necessary peak power. This clearly demonstrates the need and importance of addressing the *dynamic behavior* of the system

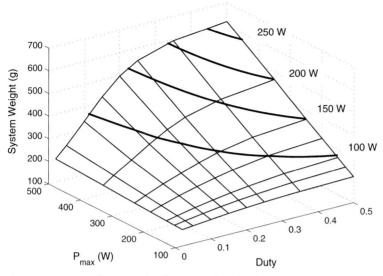

Figure 12.4 System design weights for varying duty d and maximum power P_{max} at a period of 30 s. Contours of average power are overlaid in heavy lines corresponding to values of 100, 150, 200, and 250W.

during the design phase in order to achieve fully optimal results, as explored more fully in the chapter devoted to optimal design and operation and in [29].

The system model used in the simultaneous dynamic optimization to this point is based on very simplistic phenomenological models. It is also important to demonstrate the ability to apply this optimization technique to more realistic physics-based models as well. A second system model was developed in which the lumped-capacitance battery model was replaced by an intermediate complexity lithium-ion particle model [30–32] containing seven DAEs. The ideal capacitor model is also replaced by an electrochemical capacitor porous electrode model. This model is based on work by Farahmandi [33], relying heavily on porous electrode theory [34]. The equations are fully derived in Chapter 8 of [35]. Numerical studies indicated that 10 intervals are sufficient to discretize the PDE in the capacitor model. Therefore, the electrochemical capacitor model has 11 spatial node points for potential and a 12th state for current. The fuel cell model remained the same as before. Combined together with the electronic load, the more complex model consists of 21 states in total. Due to the increased size of the system, the number of intervals in the discretization was reduced from 200 to 150. The resulting optimization problems contained 13 076 variables and 13 071 constraints. Due to the increased problem size and number of highly nonlinear terms, these problems exhibited an approximately 10-fold increase in solution time. Consequently, only a limited number of design permutations were examined.

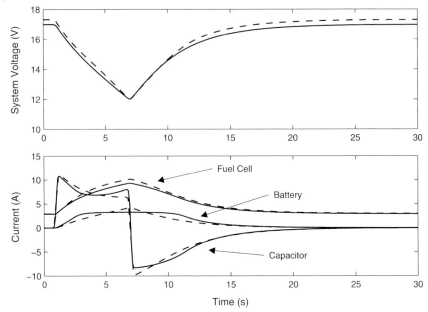

Figure 12.5 Comparison of dynamic profiles between simplified model (dashed line) and more complex physics-based model (solid line).

Figure 12.5 compares the optimal dynamic profile obtained from the complex model to that of the simple model for a pulsed-power load with a period of 30 s, duty of 20%, and a peak power of 250 W. The optimal voltage and current profiles of both systems are very similar. Moreover, the optimal parameter values and system mass are very similar, as shown in Table 12.2. After rounding the integer parameters to the next greatest value, the design sizes are nearly equivalent. These results convey two important messages: (i) the methodology and problem formulated in this example is suitable for complicated physics-based models of moderate

Table 12.2 Optimal design configurations for 30 s pulsed-power load at 20% duty and 250 W peak power. Results were obtained in under 5 min for the empirical model versus about 50 min for the physics-based model.

Parameter	Empirical model	Physics-based model
$N_{BattSer}$	4.3	4.0
$N_{BattPar}$	1.0	1.0
N_{Cap}	1.8	1.4
N_{FC}	18.7	18.2
A_{FC} (cm^2)	16.7	18.4
Mass (g)	302	291

size; and (ii) the use of trusted reduced order models can greatly reduce the necessary computation time when solving design optimizations.

Figure 12.5 shows that the battery current is consistently positive. This means that the battery is not recharging during the cycle. A soft constraint was used to force the terminal current for the battery and capacitor to 0. After a few cycles, the drop in battery state of charge would lower the open circuit potential and allow the battery to begin recharging. This should not alter the results or conclusions significantly, particularly since the number of batteries would be rounded up for any system actually implemented. Also note that the capacitor current is not an ideal step function. The steps are smoothed with a continuous approximation. The unit step is modeled by:

$$y = 0.5 \tanh(50x) + 0.5$$

which is then scaled and shifted appropriately. When using this approximation, >96% of the step occurs within 0.08 s, which should be sufficient for this study. Furthermore, the size of the sampled elements was $30/200 = 0.15$ s for the simple system and $30/150 = 0.2$ s for the more complex system. This may also contribute to the response not resembling an ideal step function.

12.5
Future Directions

There are a variety of emerging research areas that could be considered in future efforts. One area for possible development would be better coupling of complex PDE models for spatially-integrated components to more traditional component modeling tools. Obviously, the more computationally demanding PDE systems would decrease the computational speed of the system, but there may be an opportunity for automated model reduction of these PDE systems allowing direct translation to low order approximate DAE systems. The resulting simulation system may be used in black box optimization schemes when considering design issues. Without explicit knowledge of the underlying model equations, this approach would be severely limited for consideration in constrained systems. If the resulting dynamic model is of small to moderate scale, collocation methods can be used to create a viable design optimization formulation.

To complicate this problem even further, discrete design variables can be considered in the optimization problem. In the example problem of this chapter, many of the scaling variables were truly discrete, but were able to be treated as continuous in the formulation of the problem. However, many instances exist where this would not be possible. For instance, the problem could be formulated with a superstructure that contains many potential components and system elements that may or may not participate in the optimal design. This allows the engineer to answer such questions as: Do you include a lithium ion battery or a NiCd battery? Which battery model should be used? What control structure should be used to regulate this system? Clearly, a choice of one half of a lithium ion battery and one half of a NiCd battery is not a feasible solution; binary decisions must be considered explicitly. In some cases, these problems can be solved by simple enumeration. Other cases may require the use of more complicated integer optimization methods. Traditionally mixed-integer methods have been limited to problems with linear objective functions and linear constraints but rapid progress is being made in mixed-integer nonlinear optimization methods for solution of large-scale engineering problems [36–38]. Furthermore, others [39] and references therein, have presented design (or simultaneous design and control) problems as mixed-integer dynamic optimization (MIDO) problems along with methods to formulate them as MINLPs that can be solved by available methods.

One important area for the design of hybrid power systems is the consideration of active power sharing between components. In the work presented in this chapter, the components were coupled together passively using a parallel topology. Considerable work has been done in the field of power electronics to develop compact and efficient active switching DC–DC converters such as the buck (step-down) or boost (step-up) designs. This allows different components to operate at different voltages, allowing each component to operate near its maximum power output. Power converter design could be considered part of the final design task, to be completed after the components have been specified. However, there are interesting design challenges in the area of power conversion. One must determine where the power conversion device is located in the system (i.e. which component is isolated from the main system bus). One must also develop control strategies to operate the power converter. The main drawback other than complexity and cost to active power sharing using power converters is the loss of some power in the voltage conversion process, but progress is being made in new topologies and components (e.g. metal oxide semiconductor field effect transistors (MOSFETs)) that lead to a higher conversion efficiency.

One final consideration is the use of uncertainty in the formulation. Model parameters may be known within a given confidence interval or statistical distribution. The desired power profile may be only partially known. Given an expected probability distribution for parameters, the design problem can be posed as design under uncertainty. There are emerging methods available to approach these types of problems [40, 41]. Generally, methods for optimization under uncertainty require small problem formulations or rely on heuristic search methodologies.

Acknowledgments

This work has been partially supported by the National Science Foundation CAREER Award CTS-0238663 as well as by Army Research Office STTR Contract W911NF-06-C-0099. The authors would also like to acknowledge useful conversations with colleagues at the University of South Carolina, including Dr Ralph White, Dr Godfrey Sikha, Dr Roger Dougal, Dr Anto Monti, and Dr Enrico Santi. Finally, the authors would like to sincerely thank the editors of this book for very useful feedback and suggestions.

References

1 Galan, S. and Barton, P.I. (1998) Dynamic optimization of hybrid systems. *Computers & Chemical Engineering*, **22**, S183–90.

2 El-Farra, N.H., Mhaskar, P. and Christofides, P.D. (2005) Output feedback control of switched nonlinear systems using multiple lyapunov functions. *Systems & Control Letters*, **54** (12), 1163–82.

3 Burke, A.F. (1995) Title placeholder, in Proceedings of the 36th Power Sources Conference, Cheery Hill, NJ.

4 Miller, J.R. (1995) Title placeholder, in Proceedings of the Fifth International Seminar on Double Layer Capacitor and Similar Energy Storage Devices, Boca Raton, Florida.

5 Zheng, J.P., Jow, T.R. and Ding, M.S. (2005) Hybrid power sources for pulsed current applications. *IEEE Transactions on Aerospace and Electronic Systems*, **37** (1), 288–92.

6 Holland, C.E., Weidner, J.W., Dougal, R. A. and White, R.E. (2002) Experimental characterization of hybrid power systems under pulse current loads. *Journal of Power Sources*, **109** (1), 32–7.

7 Atwater, T.B. and Cygan, P.J. (1996) Title placeholder, in Proceedings of the 37th Power Sources Conference, Cheery Hill, NJ.

8 Chu, A. and Braatz, P. (2002) Comparison of commercial super-capacitors and high-power lithium-ion batteries for power-assist applications in hybrid electric vehicles I. initial characterization. *Journal of Power Sources*, **112** (1), 236–46.

9 Sikha, G. and Popov, B.N. (2004) Performance optimization of battery-capacitor hybrid system. *Journal of Power Sources*, **134**, 130–8.

10 Chandrasekaran, R., Sikha, G. and Popov, B.N. (2005) Capacity fade analysis of a battery/super capacitor hybrid and a battery under pulse loads at elevated temperature part i-full cell studies. *Journal of Applied Electrochemistry*, **35**, 1005–113.

11 Nadal, M. and Barbir, F. (1996) Development of a hybrid fuel cell/battery powered electric vehicle. *International Journal of Hydrogen Energy*, **21** (6), 497–505.

12 Jarvis, L.P., Atwater, T.B., Plichta, E.J. and Cygan, P.J. (1998) Power assisted fuel cell. *Journal of Power Sources*, **70** (2), 253–7.

13 Jiang, Z. and Dougal, R.A. (2003) Design and testing of a fuel-cell powered battery charging station. *Journal of Power Sources*, **115** (279), 95–108.

14 Kim, M. and Peng, H. (2007) Power management and desing optimization of fuel cell/battery hybrid vehicles. *Journal of Power Sources*, **165**, 819–32.

15 Dougal, R.A., Liu, S. and White, R.E. (2002) Power and life extension of battery/ultracapacitor hybrids. *IEEE Transactions on Components and Packaging Technologies*, **25** (120), 120–31.

16 Cadence Design Systems Inc. (2007) PSpice® A/D, PSpice Advanced Analysis. http://www.cadence.com/datasheets/pspice_ds.pdf (accessed 17 June 2007).

17 The MathWorks Inc. (2007) SimPowerSystems 4 User's Guide. http://www.mathworks.com/access/helpdesk/help/pdf_doc/physmod/powersys/powersys.pdf (accessed 17 June 2007).

18 Broughton, E. Langland, B. Solodovnik, E. and Croft, G. (2005) VTB User's Manual. University of South Carolina, http://vtb.ee.sc.edu/developers/VTB_Users_Manual.pdf (accessed 17 June 2007).

19 Petzold, L.R. (1982) A description of DASSL: a differential/algebraic system solver. Technical Report SAND82-8637, Sandia National Laboratories, September 1982.

20 Slater, J.C. (1934) Electronic energy bands in metals. *Physical Reviews*, **45**, 794–801.

21 Lapidus, L. and Pinder, G.F. (1982) *Numerical Solution of Partial Differential Equations in Science and Engineering*, John Wiley & Sons, Inc., New York.

22 Davis, M.E. (1984) *Numerical Methods and Modeling for Chemical Engineers*, John Wiley & Sons, Inc., New York.

23 Rice, R.G. and Do, D.D. (1995) *Applied Mathematics and Modeling for Chemical Engineers*, John Wiley & Sons, Inc., New York.

24 Kameswaran, S. and Biegler, L.T. (2006) Simultaneous dynamic optimization strategies: recent advances and challenges. *Computers & Chemical Engineering*, **30**, 1560–75.

25 Kameswaran, S. and Biegler, L.T. (2007) Convergence Rates for Direct Transcription of Optimal Control Problems using Collocation at Radou Points. *Computational Optimization and Applications*, **41** (1), 81–126.

26 Stamps, A.T., Holland, C.E., White, R.E. and Gatzke, E.P. (2005) Analysis of capacity fade in a lithium ion battery. *Journal of Power Systems*, **150**, 229–39.

27 Kim, J., Lee, S.-M., Srinivasan, S. and Chamberlin, C.E. (1995) Modeling of proton exchange membrane fuel cell performance with an empirical equation. *Journal of the Electrochemical Society*, **142** (8), 2670–4.

28 Wächter, A. (2002) An Interior Point Algorithm for Large-Scale Nonlinear Optimization with Applications in Process Engineering. PhD thesis, Carnegie Mellon University, January, 2002.

29 Yunt, M., Chachuat, B., Mitsos, A. and Barton, P.I. (2008) Designing man-portable power generation systems for varying power demand. *AIChE Journal*, **54** (5), 1254–69.

30 Haran, B.S., Popov, B.N. and White, R.E. (1998) Determination of the hydrogen diffusion coefficient in metal hydrides by impedance spectroscopy. *Journal of Power Sources*, **75** (1), 56–63.

31 Ning, G. and Popov, B.N. (2004) Cycle life modeling of lithium-ion batteries. *Journal of the Electrochemical Society*, **151** (10), A1584–91.

32 Santhanagopalan, S., Guo, Q., Ramadass, P. and White, R.E. (2006) Review of models for predicting the cycling

performance of lithium ion batteries. *Journal of Power Sources*, **156**, 620–8.

33 Farahmandi, C.J. (1996) A mathematical model of an electrochemical capacitor with porous electrodes, in *The Electrochemical Society Proceedings Series*, Vol. **96–25** (eds F.M. Delnick, D. Ingersoll, X. Andrieu and K. Naoi), The Electrochemical Society, Pennington, NJ, pp. 167–79.

34 Newman, J. and Thomas-Alyea, K.E. (2004) *Electrochemical Systems*, 3rd edn, Wiley-Interscience.

35 Stamps, A.T. (2007) Techniques for Parameter Estimation, Simulation, and Optimization of Dynamic Electrochemical Systems. PhD thesis, University of South Carolina.

36 Smith, E.M.B. and Constantinos, C. (1997) Pantelides. Global optimisation of nonconvex MINLPs. *Computers & Chemical Engineering*, **21** (Suppl.), S791–6.

37 Sahinidis, N.V. and Tawarmalani, M. (2005) BARON 7.2.5: Global Optimization of Mixed-Integer Nonlinear Programs, User's manual.

38 Kesavan, P., Allgor, R.J., Gatzke, E.P. and Barton, P.I. (2004) Outer approximation algorithms for separable nonconvex mixed-integer nonlinear programs. *Mathematical Programming*, **100**, 517–35.

39 Flores-Tlacuahuac, A. and Biegler, L.T. (2007) Simultaneous mixed-integer dynamic optimization for integrated design and control. *Computers & Chemical Engineering*, **31**, 588–600.

40 Acevedo, J. and Pistikipoulos, E.N. (1996) A parametric MINLP algorithm for process synthesis problems under uncertainty. *Industrial and Engineering Chemistry Research*, **35** (1), 147–58.

41 Sahinidis, N.V. (2003) Optimization under uncertainty: state-of-the-art and opportunities, Proceedings of FOCAPO.

13
Control of Microreactors

Mayuresh V. Kothare

13.1
Introduction

For the purpose of this chapter[1], following [8], we define a microchemical system as having the following characteristics:

1. It carries out chemical transformations (reactions) and/or separations;
2. it is fabricated using MEMS microfabrication methods;
3. it uses silicon and related integrated circuit (IC) industry materials but can include other materials such as polymers, ceramics, glass, quartz, and so on;
4. it contains microfluidic and non-electronic feature sizes in the range of sub-microns to a few hundred microns;
5. it integrates non-electronic features with at least one electronic feature, for example a resistive heater;
6. its main function is chemical or electrochemical *synthesis* as opposed to *analysis* or sensing.

To facilitate the discussion, we consider the schematic of a prototypical integrated microchemical system shown in Figure 13.1. The microplant integrates classical chemical unit operations at a microscale: a mixing/heating unit, a catalytic microreactor and a membrane microreactor/microseparator. The overall goal of this microplant is to produce high purity product C from liquid phase reactants A and B. The vaporizer serves to perform phase transformation and homogenization of A and B. The catalytic microreactor serves as the central unit for carrying out the heterogeneous gas phase reaction of A and B to produce product C and an undesirable and possibly hazardous byproduct D. The membrane microreactor serves to convert hazardous product D to a more benign waste product E while simultaneously separating the desired product C from unreacted A, B and byproducts D, E. Integrated resistive temperature sensors measure inlet and outlet temperatures of the microreactor and feed back a voltage signal to the controller module. The controller in turn sends appropriate currents to the resistive heaters

---

1) Portions of this chapter are excerpts from Refs [1–7].

Microfabricated Power Generation Devices. Edited by Alexander Mitsos and Paul I. Barton
Copyright © 2009 WILEY-VCH Verlag GmbH & Co. KGaA, Weinheim
ISBN: 978-3-527-32081-3

which provide heat input to the vaporizer and the microreactor to control the two temperatures.

The cross-sections of the individual micro-units, shown in Figure 13.1, indicate that the entire microplant is housed in microchannels fabricated in a silicon substrate and capped with an appropriate base plate. These structures are fabricated using standard surface and bulk micromachining techniques [9] such as photolithography, pattern transfer, sputtering, chemical vapor deposition, alkaline etching and plasma etching. While microfabrication of such a microchemical system prototype is a research problem in its own right, our emphasis in this chapter will be primarily on the dynamical analysis of the fabricated microsystem.

The prototype shown in Figure 13.1 is abstract and captures the essence of a real microchemical system, as defined above, while serving as a test-bed for formulating a number of relevant system theoretic problems in a general microchemical system. Furthermore, it generalizes a number of potential reaction/ separation/purification schemes from classical chemical engineering as well as electrochemical reaction schemes involving fuels and oxidants which can be tested in a microchemical setting. Examples include the classical hydrocarbon reforming–shift–reforming reaction sequence with or without membranes

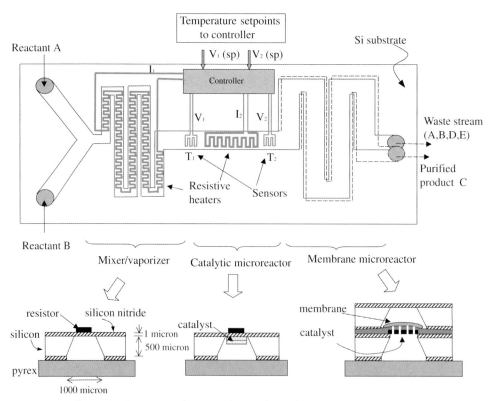

Figure 13.1 A prototypical integrated microchemical system.

[10–14], methanol dehydrogenation to formaldehyde [15] and dehydrogenation of cyclohexane to benzene [16].

The operational goals of the prototype may be considered to be one or more of the following:

1. Stabilization of the overall microplant at the chosen operating conditions;
2. maximization of the microsystem throughput;
3. maximization of conversion of the reactants to the desired product C;
4. minimization of the amount of hazardous byproduct D in the waste stream;
5. satisfaction of constraints on key process variables to ensure integrity of the microsystem material.

Inherent in such a problem formulation is the need to (i) understand the impact of the microsystem design (shape, size, length of microchannels, microunit topology) on its performance; (ii) develop and analyze models that can be used as the basis for microreactor design, operational optimization and multi-unit feedback controller synthesis; and (iii) study the robustness of operation, that is understand the effect of microfabrication errors [17, 18], imperfections in microreactor geometry [19] and structural/parametric model uncertainty on closed-loop controller performance. It is worth noting that the aforementioned set of problems is independent of the specific microsystem geometry, configuration and reaction kinetics and, therefore, is not limited to the specific prototype shown in Figure 13.1.

Our goal in this chapter is to consider generic issues that arise when dealing with control problems for microsystems. While controller design for micropower systems is not specifically addressed, the generic discussion provides insight into the considerations involved when addressing controller formulation and implementation for micropower systems.

13.2
Issues in Control of Microsystems

The study of the dynamical properties and feedback control of microreactors poses unique challenges. These challenges stem from the unique characteristics of microsystems, namely, their small size, high surface area to volume ratios implying higher heat and mass transfer rates, low thermal inertia, fast transients, and small available physical space for building and incorporating the controller implementation with the microreactor system.

A variety of heuristics from macroscale systems that allow considerable decoupling of control loops become inapplicable in the context of microreactors due to very strong integration of various unit operations fabricated in close proximity to each other on a common substrate. Similarly, the notion of controller "implementation", which is typically relegated to "computer" control for macrosystems with appropriate data acquisition and feedback signal transfer, is no longer simple for microchemical systems since an integrated microchemical system must also integrate a "small" controller within the microscale space constraints available. In other words, one simply cannot control a microreactor system with a computer, but the control algorithm must be "embedded" within the microreactor substrate. And finally, within the context of micro-power chemical systems, the controller implementation must minimize its parasitic power requirements to a small fraction of the total power projected from the device.

Within the context of control problems in micro and nanosystems in general, two workshops were organized by the U.S. National Science Foundation in 2003 and 2004 [20–22]. The resulting workshop reports provided a variety of recommendations on future research opportunities in the control of micro- and nano-systems. The scope of both workshops was broader than just control of microreactors and covered not only control of micro- and nano-scale devices but also control of micro/nano-fabrication processing technologies, biomimetic control and micro/nano-scale sensing that included control of AFM probes [22].

However, several of the key recommendations from these workshops are also relevant within the context of integrated microreactors and are listed below [20, 21]:

- Characterization of the impact of micro/nano-component integration on control;

- development of both fundamental multiscale models, as well as parsimonious control relevant models obtained from system identification experiments or model reduction;

- on-chip or embedded control algorithms integrated with the micro/ nano-system;

- evaluation of robustness of closed-loop in the presence of fabrication uncertainty and disturbances.

In the rest of the chapter, we will focus on issues involved in the development and implementation of model-based predictive controllers for microchemical systems.

13.3
Control Relevant Modeling of Microchemical Systems

13.3.1
Continuum Models

Fundamental model development for microchemical systems of the kind discussed in this paper poses unique challenges that stem from the complex coupling of multiple physical phenomena and the unusual geometries of these micro-devices. The complex coupling arises primarily from the strong interaction between electrical, mechanical, thermal, microfluidic and chemical phenomena in compactly configured micro-geometries. The individual phenomena by themselves are complex, particularly due to the small channel dimensions.

Flow in micro-devices is characterized by departure, to varying degree, from the continuum assumption and this is measured by a dimensionless group called the Knudsen number (*Kn*) [23]. For most microfluidic flows, *Kn* is less than 0.1 which puts the flow in the slip regime. In this regime, fluid flow can be modeled using the continuum conservative equations, with a modified boundary condition to account for wall velocity slip [24]. Flow in microchannels is predominantly laminar [25] due to the small hydraulic diameter. A number of related issues in micro-fluidic mixing and models to make predictions for improved mixing can be found in [19, 26, 27] and the references therein. A general consensus seems to be that for a large number of microchemical systems with dimensions in the size range of ten to hundreds of microns and atmospheric pressure, a purely continuum approach appears to be adequate [28].

However, the modeling problem becomes considerably more difficult when microkinetic models that incorporate atomistic scale details of chemical reactions are coupled with transport models. Typically, this approach leads to coupling of multiple time and spatial scales and is discussed in greater detail in other parts of this book. The important issue is that the computational effort required to solve even the simplest such models would be prohibitively large, making such models difficult to use in a control context. Moreover, recent work [29, 30] has shown that both continuous time Monte Carlo simulations [30] and molecular approaches to the solutions of the Navier–Stokes equation [29] were in excellent agreement with continuum solutions for Knudsen numbers below 0.1, which is typically the case for microchemical systems. For reacting systems, we adopt the approach that the microkinetic modeling leads to a lumped reaction rate equation which can then be coupled with continuum solutions for the transport equations.

Thus, with the assumptions posed above, integrated microchemical systems can be characterized by distributed parameter system (DPS) models.

Thus, control of a microchemical system can now be posed within the context of control of infinite dimensional partial differential equations, typically with boundary actuation, as discussed in the next section.

13.3.2
Model Reduction

Feedback control of microreactors by directly exploiting their distributed parameter model descriptions remains an open problem. Model reduction plays a key role in making the control problem solvable. Proper orthogonal decomposition (POD) is a widely used technique that uses second-order statistical properties to compute empirical eigenfunctions from a collection of data. In the context of DPS models for microchemical systems, the method of snapshots [31] combined with the weighted residual method (WRM) [32] can be used to develop finite dimensional low-order models.

In the above strategy, the first-principles DPS models can be used to generate snapshots or datasets using a finite element method solver such as FEMLAB (now called COMSOL Multiphysics) [33]. These snapshots can then be used to compute

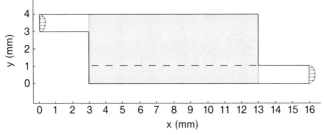

Figure 13.2 Geometry examined [3].

the dominant spatial and/or temporal eigenfunctions [3] for use in a closed-loop control strategy.

The key property of POD that makes it attractive for model reduction is that it provides the optimal basis functions for capturing the dominant features (or more precisely, energy, quantified in terms of eigenvalues) from snapshots. In other words, there is no other set of N modes that contains more energy than the first N POD eigenfunctions. As a result, several references can be found in the literature that use variations of POD in model reduction, data analysis and feedback control. Kevrikidis and coworkers [34, 35] report its use for nonlinear model reduction and data analysis. In the context of parabolic PDEs, the method of empirical eigenfunctions was used in [36] for low order system approximation and nonlinear controller design. The method has found applicability in the optimal control of turbulent fluid flows [37].

Due to its broad applicability, POD is known by different names, depending on its application domain [38]. For instance, in the pattern recognition community, it is commonly referred to as principal component analysis (PCA); in image processing, it is known as the Hotelling transform, and also the Karhunen–Loéve decomposition [39].

Within the context of microchemical systems, the use of POD to capture dominant modes of a flow problem in a microfluidic geometry (Figure 13.2) can be found in [3]. It was found in [3] that about 4 or 5 spatial or temporal eigenfunctions were sufficient to fully reproduce the spatio-temporal evolution of the flow profile in a microfluidic geometry. This is illustrated in Figure 13.3 for the flow geometry shown in Figure 13.2.

The use of these eigenfunctions in a closed-loop feedback control formulation is discussed in Section 13.4.

13.3.3
Empirical Models

We recognize that certain microchemical units may not be easily amenable to a first-principles mathematical model. For example, in the prototype shown in Figure 13.1, modeling of the mixing/vaporization step involves a phase change from fully liquid to a two-phase flow and, ultimately, a fully vaporized flow. Model-

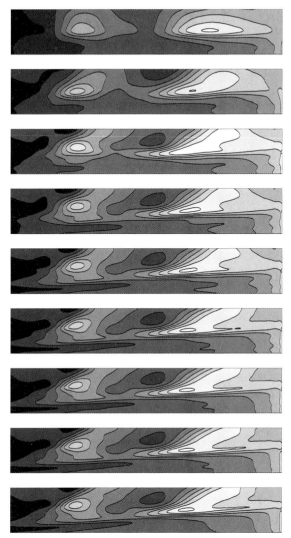

Figure 13.3 Reconstructed data ensemble using the spatial eigenfunctions. The y-axis is time, the x-axis is spatial location for the geometry in Figure 13.2. The various colors represent equal velocity contours. The top subplot is the reconstruction using one eigenfunction, the second subplot from the top is the reconstruction using two eigenfunctions and so on. The bottom is the initial data ensemble generated from the solution of the Navier–Stokes equation using FEMLAB [33] (now called COMSOL Multiphysics) for the geometry in Figure 13.2 [3].

ing of such phenomena is clearly beyond the scope of this chapter, and indeed defines a research problem in its own right [37, 40–43]. For situations such as these, controller design can be made feasible with experimental input–output data-driven models. The motivation for this approach stems from the fact that in many feedback control applications, an input–output data-driven model obtained from on-line experiments on the system suffices in designing model-based controllers. This empirical model-based control approach has a long history of successful application and acceptance in numerous control applications, in particular, those in the process industries. Similarly, by determining local models at multiple operating points, one is able to obtain a family of local models which can then be interpolated to generate a global model [44–47]. Such an approach is particularly well-suited for multi-model adaptive control techniques [46, 47].

The use of empirical models to control flow distribution in microchannels is discussed in Section 13.4.

13.4
Feedback Control of Microchemical Systems

The earliest reported work on control in MEMS devices can be found in the context of MEMS gyroscopes [48]. Empirical models identified from input–output data [18] were used to study the frequency response characteristics of a MEMS gyroscope for the purpose of controller design. In an alternative approach [49], kinematic force balance models were used to study adaptive control of a Z-axis gyroscope. Microchemical systems share several of the characteristics of MEMS devices in general and, additionally, incorporate more functionality through chemical reaction, separation and fluid flow. The coupling of the effects of heat, mass, fluid transport with chemical reaction, in conjunction with electronic sensing and actuation in a confined geometry makes control of microchemical systems substantially more involved.

Consider Figure 13.4 which shows a straight channel microreactor cross-section, with flow direction as indicated. This configuration could be a subsystem of the

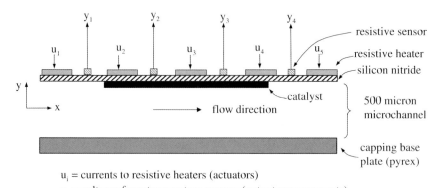

u_i = currents to resistive heaters (actuators)
y_i = voltages from temperature sensors (output measurements)

Figure 13.4 Control problem specification for a prototype microreactor.

generic microreactor shown earlier in Figure 13.1. Thin film resistive sensing and heating elements are shown on the top wall of the microchannel. We note that the sensors and actuators are *distributed* as a function of the *x*-axis, but are *boundary* elements with respect to the *y*-axis. Similarly, the system states such as temperature and species concentration are functions of space and time. Thus, this is a distributed parameter system with combined distributed and boundary sensing and actuation. We note that the distinguishing feature of the microsystem shown is that it is inherently distributed, as opposed to a number of recently reported studies on systems that are distributed because they are comprised of distributed interacting systems that arise, among other reasons, due to the ready and low-cost availability of distributed sensors and actuators.

While considerable literature exists on the control of distributed parameter systems, much of this literature does not strictly apply to the class of control problems discussed above because the actuation is at the boundary. "Boundary control" problems have only recently been addressed in the control literature [50–52]. Formulating optimal control problems within the framework of boundary-actuated distributed systems is particularly difficult. Prior work on boundary control has been in heat conduction problems [53, 54]; boundary modal control of melt temperature in crystal growth [55] and sliding mode control of an isothermal tubular reactor [56]. Identification of distributed models using boundary excitation inputs was presented in [57].

While we will not attempt to address the entire gamut of problems in boundary control, we will present two promising approaches to formulate computationally tractable controller designs for microsystems that have generic utility.

13.4.1
POD-Based Boundary Control of Thermal Transients in Microsystems

We will present a very brief overview of POD that employs the method of snapshots. Detailed derivations can be found in a variety of sources [3, 4, 58].

Suppose we have an ensemble of scalar functions $y_t(x)$ collected either experimentally or from FEM simulations of the governing partial differential equations, where x is the spatial variable on a domain $\Omega_s \in \Re$ and t is the temporal variable on a domain $\Omega_t \in \Re$. Thus we have N observations (called snapshots) of some physical process taken at positions x, where $x = 1, \dots, M$.

The goal of POD is to develop an approximate representation of $y_t(x)$ as follows:

$$y_t(x) \approx \sum_{k=1}^{N} \varphi_k(x)\psi_k(t) \tag{13.1}$$

where $\varphi_t(x)$ are called the spatial eigenfunctions of the ensemble of snapshots $y_t(x)$. The empirically determined spatial eigenfunctions $\varphi_k(x)$ are computed as follows (see [4]):

$$\varphi_k(x) = \sum_{t=1}^{N} A_t^{(k)} \gamma_t(x) \tag{13.2}$$

where $A_t^{(k)}$ are the elements of the eigenvector corresponding to the largest eigenvalue of the following eigenvalue problem:

$$C_s A^{(k)} = \lambda_k A^{(k)} \quad k = 1, \dots, N \tag{13.3}$$

where

$$(C_S)_{ij} = \frac{1}{N} \sum_{x=1}^{M} \gamma_i(x) \gamma_j(x) \quad i, j = 1, \dots, N \tag{13.4}$$

13.4.1.1 Wafer Temperature Profile Control

Consider Figure 13.5 which shows the cross-section of a wafer geometry, with thin film resistive heaters placed on the top wall and sensors at the bottom. The wafer states such as temperature are functions of space and time. Thus, this is a distributed parameter system with combined distributed and boundary sensing and actuation.

The temperature distribution in the wafer is described by the time-dependent two-dimensional heat equation

$$\frac{\partial T(t, x, y)}{\partial t} = \frac{1}{a} \nabla^2 T(t, x, y) \tag{13.5}$$

with convective boundary conditions at the top, bottom and edge surfaces of the wafer. Our control objective is to be able to maintain a desired temperature profile at the sensor locations by using the boundary-actuated heat inputs.

Our proposed approach is to use FEM simulations of the heat equations to develop snapshots of the temperatures at the sensor locations, apply POD to get the dominant spatial eigenfunction and then change the heat inputs to the boundary heaters so as to steer this spatial eigenfunction to a desired spatial eigenfunction. The desired spatial eigenfunction is in turn computed by performing a POD of the desired temperature profile at the sensor locations.

The following objective function is used to formulate this minimization problem:

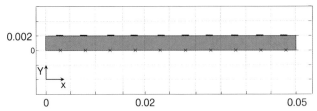

Figure 13.5 Cross-section geometry with sensor locations indicated with the "x" symbol (dimensions in meters).

$$\min_{u_t, u_{t+1}, \ldots, u_{t+N}} J_S(u) = \sum_{x=1}^{M} \left| \varphi_1(x) - \varphi_1^d(x) \right| \tag{13.6}$$

where u_t is the vector of heat inputs at the 10 heater locations at time t. Thus, this objective tries to find the optimal heat inputs at the current discrete time t and future times $t + 1, t + 2, \ldots t + N$ so as to reach the desired temperature profile at the sensors.

Inherent in this formulation is the assumption that the first eigenfunction captures essentially all the dominant spatiotemporal characteristics of the problem. A second assumption is that the magnitude of the heat inputs is small enough that the dominant eigenvector remains invariant. Both these assumptions have been verified through simulations reported in [4].

13.4.1.2 Receding Horizon Formulation
The receding horizon formulation of this POD-based control problem is shown in Figure 13.6. The algorithm combines classical receding horizon control ideas [59] with POD and is summarized below (see [4]):

13.4.1.3 Receding Horizon Control Using POD [4]
Select the tolerance $\varepsilon > 0$, control horizon N, compute desired spatial and temporal dominant eigenfunctions and perform the following steps:

Step 1: Give initial guess for the current and future manipulated inputs u_t, ..., $u_{t + N(x)}$ to the COMSOL model.
Step 2: Compute snapshots and associated spatial dominant eigenfunctions for the data ensemble $y_t, \ldots, y_{t + N(x)}$ obtained by the COMSOL model.
Step 3: Compute cost functions of (13.6). If the cost is less than ε go to Step 5.
Step 4: Adjust the boundary actuation $u_t, \ldots, u_{t + N(x)}$ under the constraint of invariant eigenvectors until both cost functions are less than ε. If the eigenvectors are

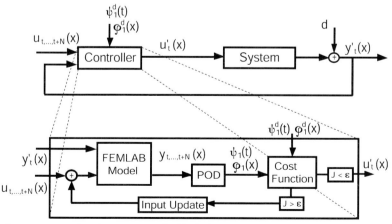

Figure 13.6 Receding horizon controller block diagram [4].

not sufficiently invariant (as measured by the tan2θ theorem [4]), decrease the boundary actuation.

Step 5: Implement the first computed input $u_t(x)$ of the computed boundary actuation on the actual system, move the control horizon one sample ahead and go to Step 2. Using feedback, the measured output of the system $y_t(x)$ is used for the initialization of the COMSOL model.

13.4.1.4 Simulation Results

We would like to track the spatiotemporal profile shown in Figure 13.7 at the 10 temperature sensors. A POD of this profile gives us the desired spatial eigenfunctions. We choose a control horizon of 20. The boundary actuation heat inputs are set to have an upper constraint of $10\,\mathrm{W\,m^{-2}}$. As an initial guess, we provide $5\,\mathrm{W\,m^{-2}}$ heat input to the 10 heaters and we obtain temperature profiles from the FEM simulations. Subsequently, the energy supply is adjusted recursively in order to minimize the cost functions J_s. Applying the proposed receding horizon algorithm we successfully control the temperature distribution both spatially and temporally, as illustrated in Figure 13.8. The corresponding manipulated heat input profile at the boundary is shown in Figure 13.9.

A variety of control problems in microchemical and micropower systems can be handled using this receding horizon approach, as long as a suitable detailed model is available for obtaining "snapshots" for carrying out POD. Similarly, within this formulation, disturbance rejection can be readily handled through

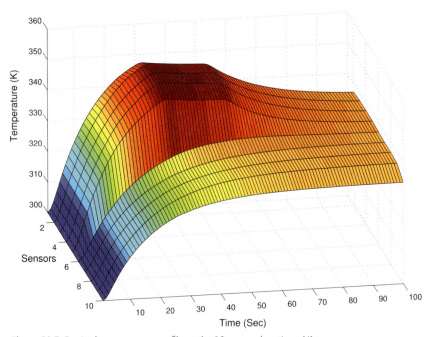

Figure 13.7 Desired temperature profile at the 10 sensor locations [4].

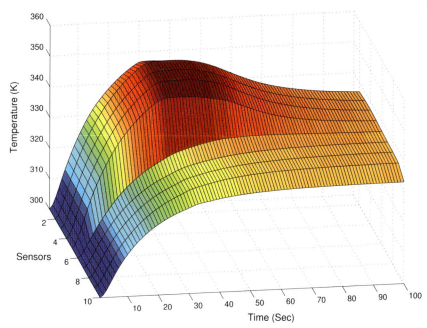

Figure 13.8 Closed-loop temperature profile for receding horizon of 20 [4].

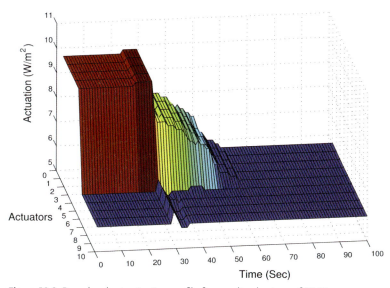

Figure 13.9 Boundary heat actuation profile for receding horizon of 20 [4].

simulation [4] and the impact of parametric uncertainty, overall robustness and the impact of design can also be evaluated.

As an example, it was shown in [4] that it is essentially impossible to maintain temperature profiles in the wafer geometry shown if the wafer material is silicon. This is due to the high thermal conductivity of silicon that very quickly equalizes temperature by conducting heat rapidly. Only low thermal conductivity materials such as ceramic, glass or silicon nitride can be used. Similarly, the COMSOL-based modeling and simulation environment also allows one to change the geometry of the microreactor and investigate the effect on control lability. Such simulation experiments were not feasible a few years ago, but are now much more routinely feasible with COMSOL.

13.4.2
Empirical Model-Based Predictive Control of Microflows

As indicated before, a variety of microchemical systems may not be easily amenable to first-principles mathematical modeling. In other cases, first-principles models, even with the POD approach described in the previous section, may be computationally far too expensive for implementation in real-time. In either of these cases, models developed from input–output data from experiments or from solution of first-principle models can be used effectively in model-based feedback control.

Consider the microfluidic geometry shown in Figure 13.10 [1]. There are three inlets (left side) and two outlets (right side). The middle inlet from the left side allows the introduction of chemicals in the microgeometry. The control task is to use the top and bottom inlet stream flow rates to dynamically steer the middle inlet stream species to either of the two outlets (switch) or distribute it appropriately between the two outlets.

Flow in microchannels is predominantly laminar [28] due to the small Reynold number (*Re*). As a consequence of the low *Re*, we can introduce two or more different fluids and expect a distinguishable interface between them because of the

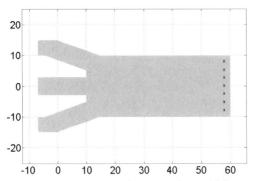

Figure 13.10 Geometry used to create the model for MPC (dimensions in microns) [1].

lack of turbulence and small molecular diffusion between the streams. Thus, the top and bottom inlet stream velocities can be used to apply hydrodynamic control to the middle inlet stream, thereby accomplishing the control task.

Liquid flow in microchannels can be reasonably described with continuum models (see [28] for details), using the continuity equation and Navier–Stokes equations for each spatial component of the velocity vector v in the x, y directions (neglecting z dependence). The boundary conditions are no-slip at the walls and full developed laminar flow. In order to simulate the dynamic behavior of the chemical species inserted in the middle inlet we assume Fickian diffusion of dilute species. To describe the diffusive transport in the microchannel we use the classical convection–diffusion equation with the Danckwerts boundary condition.

13.4.2.1 COMSOL (FEMLAB) Simulation

We utilize COMSOL [33] for modeling and simulating the resulting partial differential equations (PDEs) within the MATLAB [60] environment. A unique feature of COMSOL is its ability to export velocity and concentration profiles of the simulated system back to MATLAB, to carry out rigorous analysis of the dynamics and, most importantly, to incorporate controller designs in a closed-loop setting in the MATLAB/SIMULINK environment.

In an attempt to develop an empirical model from this detailed COMSOL simulation, we apply an impulse change of $0.05\,\mathrm{mm\,s^{-1}}$ magnitude to each of the upper and lower left flow inlets. We measure velocities and concentrations at discrete locations close to the outlets, as indicated with the symbols "x" in Figure 13.10. This then provides us with an impulse response model for predicting the future outputs $y(t + k|t)$ of the system using the finite impulse response (FIR) model form:

$$y(t+k|t) = \sum_{i=t}^{t+k} h(t+k-i)u(i|t) \tag{13.7}$$

where y is a vector with its components being the velocity and concentration at the discrete locations "x" at the outlet, u is the vector of manipulated inputs which are the inlet peak velocities at the top and bottom, and h is the impulse response coefficient. The argument $y(t + k|t)$ denotes the value of the vector y predicted at discrete time $t + k$ using information at time t. This is standard terminology used within the context of model predictive control (MPC) (see [59] for a tutorial on MPC).

This empirically identified FIR model of the coupled fluid flow and convection–diffusion problem can now be readily used within a receding horizon control or MPC framework. While in this specific case, we have developed this FIR model from simulation of the detailed first-principle model, such an FIR model can also be developed by doing input–output experiments, injecting impulse responses into an experimental system and measuring the resulting outputs. Standard software routines exist for identifying the impulse response coefficients $h(k)$ from such experimental data. Some of these routines are also available in the MATLAB environment.

13.4.2.2 Model Predictive Control (MPC)

Controllers belonging to the MPC family are generally characterized by the following steps: at each sampling instance, the future system outputs are calculated over a predetermined horizon P, called the prediction horizon, using the process model. These future predicted outputs, predicted at sampling instance t and denoted by $y(t + k|t)$ for $k = 1, 2, \ldots, P$ depend on the past inputs and on the current and future manipulated input signals $u(t + k|t)$, $k = 0, \ldots, N - 1$. Here, N is called the control horizon. MPC uses this predictive capability to calculate the set of future control moves $u(t + k|t)$, $k = 0, 1, \ldots, N - 1$ by optimizing a suitable cost criterion in order to keep the system as close as possible to a predefined future reference trajectory. This criterion is usually a quadratic function of the difference between the predicted output signal and the reference trajectory. In some cases the control moves $u(t + k|t)$ are included in the objective function in order to minimize the control effort. A typical objective function is the following quadratic form:

$$J_P(k) = \sum_{k=0}^{P} \left\{ \left[y(t + k|t) - y_{ref} \right]^2 + Ru(t + k|t)^2 \right\} \tag{13.8}$$

$$|u(t + k|t)| \leq b, \quad k \geq 0 \tag{13.9}$$

where $y(t + k|t)$ are the predicted outputs, y_{ref} is the desired setpoint or reference output, $u(t + k|t)$ the control sequence and R is the weighting on the control moves, a design parameter. This system is subject to input constraints given by the vector b.

Once the optimal control moves $u(t + k|t)$, $k = 0 \ldots N - 1$ are computed, only the first control move $u(t|t)$ is implemented by the controller on the system while the rest are rejected. At the next sampling instant $t + 1$, the control and prediction horizon are shifted forward by one step, the output $y(t + 1)$ is measured by the system and the procedure is repeated by the new measurements so that we get an updated control sequence. Since the control and prediction horizon both shift forward at each discrete time t, MPC is also sometimes called receding horizon control (RHC) or moving horizon control (MHC).

A variety of model forms (linear, time varying, nonlinear, hybrid, discrete, piecewise affine, ARMA), constraints (linear, nonlinear, mixed integer), objective functions (linear, quadratic, nonlinear) and stability criteria have been formulated and incorporated within the MPC framework. This flexibility of MPC as well as its theoretically rigorous and robust performance guarantees has made it the dominant optimal controller design paradigm in a variety of engineering disciplines.

One can readily show (see [59]) that the problem of minimizing the constrained quadratic objective function (13.8) using the predictive model (13.7) to determine the optimal control profile $u(t|t)$, $u(t + 1|t)$, ..., $u(t + N|t)$ is a standard quadratic programming (QP) problem. Thus, in essence, MPC computes the solution to this QP at each sampling time and implements the first computed move $u(t|t)$.

13.4.2.3 Simulation Results

As discussed before, the control objective is to appropriately adjust the inlet streams in order to drive the chemical species inserted in the middle inlet to the bottom outlet. The inlet streams enter the geometry with initial velocity of $0.05\,\mathrm{mm\,s^{-1}}$. The manipulated variables are constrained at a lower bound of $0.05\,\mathrm{mm\,s^{-1}}$ and an upper bound of $0.5\,\mathrm{mm\,s^{-1}}$. We also impose a constraint on the rate of change of the inlet velocities, by increasing the weight on control moves during the optimization.

We measure the velocities and concentrations at the seven assigned locations close to the outlets (Figure 13.10). The optimal performance for MPC is achieved using a control horizon of 6 and a prediction horizon of 10. We track the flow of the chemical species that are inserted in the microfluidic geometry, using massless particles as tags, and we examine their spatiotemporal behavior. As illustrated in Figure 13.11, the model predictive controller steers the particles inserted in the middle inlet to the bottom outlet, thus meeting the control objective within 30 s without violating any constraints.

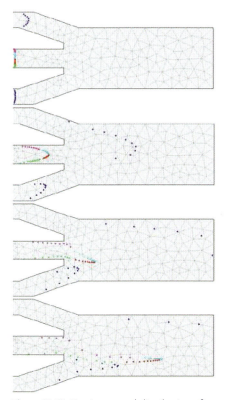

Figure 13.11 Spatiotemporal distribution of massless particle tags to track the flow of the inserted chemical species [1].

13.5
Hardware Embedded Model Predictive Control for Microchemical Systems

As we discussed earlier in the chapter, the criteria that the final controller implementation must satisfy in the context of microchemical and micropower systems are summarized below:

- The controller must be small enough to be "embedded" within the physical system, that is the microreactor.

- The controller response time must be small, that is of the order of milliseconds, to respond to potentially fast transients in microchemical and micro-combustion systems.

- The parasitic power requirements of the controller must be minimal to make its implementation viable in the context of micropower systems.

- The controller cost should be preferably, but not necessarily, a small component of the overall device cost.

- The controller must be appropriately thermally and chemically isolated from the microreactor system to protect the temperature-sensitive CMOS circuitry.

Standard single-input single-output (SISO) controllers such as the proportional-integral-derivative (PID) can certainly be readily embedded in hardware while meeting all the requirements posed above. However, when considering model-based optimal controllers, satisfying the above requirements requires more careful attention.

We will describe two approaches that we have studied to address this problem. One involves the use of off-the-shelf hardware platforms for implementing the arithmetic operations involved in a predictive controller within a small size processor. The second approach involves customizing the hardware with the purpose of optimizing its size, speed and power consumption in executing the arithmetic operations in MPC.

13.5.1
Embedded MPC on a Motorola Processor [2, 61]

Figure 13.12 shows the Motorola's high performance phyCORE-MPC555 board (no relation to the acronym MPC in this chapter). This board packs the power of Motorola's embedded 32-bit MPC555 microcontroller within a miniature footprint. The actual processor itself is only about 2×2 cm and the remaining components on the board are for signal acquisition, communication, and so on. The MPC555 is a high-speed 32-bit central processing unit that contains a 64-bit floating point unit designed to accelerate the advanced algorithms necessary to support complex applications. All signals and ports of the MPC555 extend to two Molex high density (0.635 mm pitch) 160 pin header connectors, allowing it to be plugged into user target hardware.

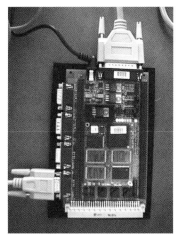

Figure 13.12 PhyCORE-MPC555 board with the MPC555 processor (2×2 cm). MPC555 has no relation to the acronym MPC used in this chapter [2, 61].

As reported in [2, 61], in order to implement the optimization problem in MPC on this target, the Code Warrior integrated development environment (IDE) [62] was used in conjunction with MATLAB/SIMULINK real-time workshop to compile a C++ implementation of the optimization routine and this was downloaded on the MPC555 processor. The size of the code downloaded was on average 30 KBytes. The phyCORE-MPC555 can be populated with a maximum of 4 MB flow-through synchronous external BURST-SRAM for data storage and a maximum of 4 MB Flash-ROM memory at 0 Wait-States.

Figure 13.13 shows the board connected to a host workstation that runs a Simulink plant model in non-real time. The code running on the MPC555 exchanges signals via RS232 serial communication with the Simulink simulations. At each sample time, Simulink performs model updates and sends an output signal via RS232 to the MPC555.

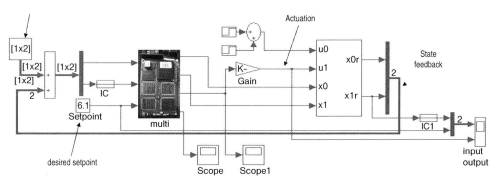

Figure 13.13 Block diagram of the embedded MPC algorithm on the Motorola board in closed loop with the SIMULINK model [2, 61].

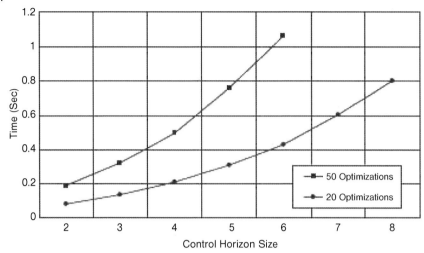

Figure 13.14 Profiling results for different control horizons with 20 and 50 optimization iterations at each sampling time [61].

The computational performance of the embedded MPC running on the MPC555 processor was studied by varying the control horizon and also by varying the number of iterations of the QP performed at each sampling time. The effect of control horizon on computation cost is given in Figure 13.14 and the effect of the number of iterations is shown in Figure 13.15. As expected, Figure 13.15 shows a linear growth while the control horizon has a stronger impact on computation

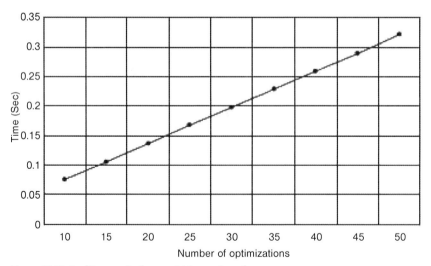

Figure 13.15 Profiling results for a variable number of optimization iterations at each time step, with fixed control horizon of 3 and prediction horizon of 10 [61].

cost. One iteration of the MPC optimization takes about 53 ms with this processor for a system with two states and one input for a control horizon of 7 and prediction horizon of 21.

13.5.2
Embedded MPC on Customized Hardware

The general purpose processor of the kind discussed in the previous section has a number of redundant units built into it to provide broader functionality that is not needed for our problems. An alternative approach to embedding MPC computations is to customize the hardware by suitable partitioning of the arithmetic operations, thereby optimizing speed, area and power consumption [5, 7].

Hardware/software (HW/SW) codesign refers to the simultaneous consideration of hardware and software within the design process. Codesign is becoming an increasingly more important research field, primarily because it provides a compromise between the high flexibility and reduced performance offered by pure software implementations [61], and high performance but reduced flexibility offered by pure hardware implementations [63, 64]. A general codesign flow is depicted in Figure 13.16.

Initially, the specifications are set and the software/hardware partitioning follows. The decision on the appropriate partitioning is based on a profiling study

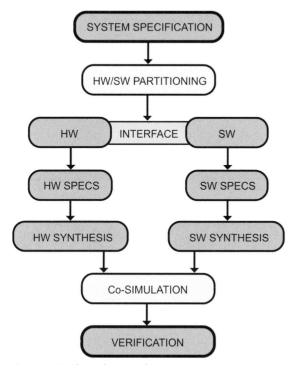

Figure 13.16 The codesign path.

of MPC, which helps to identify the computationally demanding parts of the algorithm. The computationally intensive parts of the MPC algorithm are migrated to a matrix co-processor ("hardware") while the rest of the algorithm is hosted by the general purpose microprocessor ("software") which also performs higher level operations and algorithmic control.

Figure 13.17 shows the results of a typical profiling study on a Pentium processor to identify which operations in a typical MPC algorithm are computationally the most intensive. The figure shows the percentage of time spent on each of the blocks of the algorithm for a fixed prediction horizon of 20 and a variable control horizon, for a benchmark problem. It is apparent that except for the case of a control horizon of 2, the Gauss–Jordan inversion is the main bottleneck of the algorithm and the Hessian computation becomes more time-consuming at larger controller horizons. This clearly gives an indication of the suggested software/hardware partitioning.

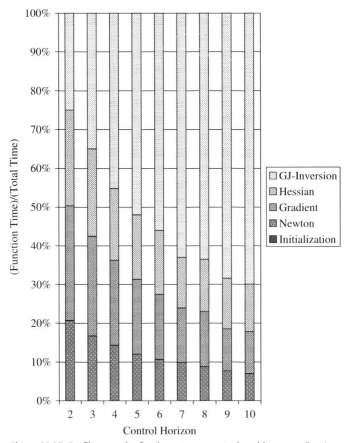

Figure 13.17 Profiling results for the antenna-control problem on a Pentium processor.

After the communication protocol between the two parts is specified, these are implemented by using a Hardware-Description Language (HDL) for the hardware and a high-level-programming language for the software. The next step is to co-simulate the two parts in order to verify the correct functionality and the performance of the complete design. This is typically done using a Field Programmable Gate Array (FPGA), the XC4VLX25-FF668-10C Virtex- IV FPGA of Xilinx being one such choice reported in [7]. If the verification process fails, the design iteration is repeated by selecting a different hardware/software partitioning.

The standardized arithmetic system for real-number arithmetic is the FP. Nevertheless, the profiling results are a clear indicator that alternative design options may be a better fit for an MPC on-chip implementation. The logarithmic number system (LNS) has appeared as an alternative to FP arithmetic [65], and recent studies prove that for word lengths up to 32-bits, LNS is more efficient than FP, with an increasing efficiency as the word length decreases [66]. In LNS, a real number is represented by the logarithm of its absolute value with an additional bit representing the sign of the real number. Due to the logarithmic representation, in LNS, the calculation of the square, cube and inverse of a real number can be implemented very efficiently, since they are reduced down to the simple operations of left shifts, additions and two's complement conversions, with simpler, faster and less power-hungry circuits than the ones in an equivalent FP unit.

In [7], the matrix co-processor ("hardware") portion of the hardware/software co-design is built using LNS, thereby providing considerable acceleration in performance. A more detailed discussion of the architecture of this LNS based co-processor coupled with a microprocessor in closed-loop with Matlab/Simulink can be found in [7]. But the key result is that for a clock frequency of 25 MHz, the computation for one optimization iteration of MPC for a system with two states and one input is about 0.89 ms for a prediction horizon of 20 and control horizon of 3. A similar size problem implemented on Motorola's phyCORE-MPC555 board takes about 15 ms per optimization iteration.

13.6
Conclusions

In this chapter, we have presented an overview of the challenges involved in formulating control problems for microchemical systems. While we have not specifically addressed control problems for micropower systems, this chapter serves to shed light on the issues that need to be addressed when reacting systems are small and in confined geometries. We have discussed several approaches to addressing these issues, both at a theoretical level and at a practical and implementation level. These generic approaches should provide fruitful starting points for formulating meaningful strategies for feedback control of micropower systems, while at the same time providing insight into the impact of systems design and materials of construction on control, and the impact of final implementation constraints on the choice of the control algorithm.

References

1 Bleris, L.G., Garcia, J.G., Arnold, M.G. and Kothare, M.V. (2006) Model predictive hydrodynamic regulation of microflows. *Journal Of Micromechanics & Microengineering*, **16** (9), 1792–9.

2 Bleris, L.G. and Kothare, M.V. (2005) Implementation of model predictive control for glucose regulation using a general purpose microprocessor, in Proceedings of the 44th IEEE Conference on Decision and Control and European Control Conference, 7, IEEE Computer Society, Seville, Spain, pp. 5162–77.

3 Bleris, L.G. and Kothare, M.V. (2005) Low order empirical modeling of distributed parameter systems using temporal and spatial eigenfunctions. *Computers & Chemical Engineering*, **29** (4), 817–27.

4 Bleris, L.G. and Kothare, M.V. (2005) Reduced order distributed boundary control of thermal transients in microsystems via the use of empirical eigenfunctions. *IEEE Transactions on Control Systems Technology*, **13** (6), 853–67.

5 Bleris, L.G., Vouzis, P.D., Arnold, M.G. and Kothare, M.V. (2007) Pathways for optimization-based drug delivery. *Control Engineering Practice*, **15** (10), 1280–91.

6 Kothare, M.V. (2006) Dynamics and control of integrated microchemical systems with application to microscale fuel reforming. *Computers & Chemical Engineering*, **30** (10–12), 1725–34.

7 Vouzis, P.D., Bleris, L.G., Arnold, M.G. and Kothare, M.V. (2009) A system-on-a-chip implementation for embedded real-time model predictive control. *IEEE Transactions on Control System. Technology*, **17** (1), (to appear).

8 Fedkiw, P., Jensen, K.F., Nowak, R. and Paur, R. (organizers) (1999) *Workshop on Microchemical Systems and Their Applications*, US Army Research Office and Defense Advanced Research Projects Agency (DARPA), Reston, VA.

9 Madou, M.J. (2002) *Fundamentals of Microfabrication: the Science of Miniaturization*, 2nd edn, CRC Press, Boca Raton, FL.

10 Franz, A.J., Schmidt, M.A. and Jensen, K.F. (1999) Palladium membrane microreactors, in The 3rd International Conference on Microreaction Technology, DECHEMA.

11 Karnik, S.V., Hatalis, M.K. and Kothare, M.V. (2003) Towards a palladium micromembrane for the water gas shift reaction: microfabrication approach and hydrogen purification results. *Journal of Microelectromechanical Systems*, **12** (1), 93–100.

12 Pattekar, A.V. and Kothare, M.V. (2004) A microreactor for hydrogen production in micro-fuel cell applications. *Journal of Microelectromechanical Systems*, **13** (1), 7–18.

13 Pattekar, A.V. and Kothare, M.V. (2005) A radial microfluidic fuel processor. *Journal of Power Sources*, **147** (1–2), 116–27.

14 Wilhite, B.A., Schmidt, M.A. and Jensen, K.F. (2004) Palladium-based micromembranes for hydrogen separation: device performance and chemical stability. *Industrial & Engineering Chemistry Research*, **43** (22), 7083–91.

15 Maurer, R., Claivaz, C., Fichtner, M., Schubet, K. and Renken, R. (2000) A microstructured reactor system for the methanol dehydrogenation to water free formaldehyde, in IMRET 4: 4th International Conference on Microreaction Technology, AIChE Spring National Meeting, Atlanta, GA, pp. 100–5.

16 Cui, T., Fang, J., Maxwell, J., Gardner, J., Besser, R. and Elmore, B. (2000) Micromachining of microreactor for dehydrogenation of cyclohexane to benzene, in IMRET 4: 4th International Conference on Microreaction Technology, Atlanta, GA, pp. 488–92.

17 Grayver, E. and M'Closkey, R.T. (2001) Automatic gain control ASIC for MEMS gyro applications, in Proceedings of the 2001 American Control Conference, Arlington, VA, pp. 1219–22.

18 M'Closkey, R.T., Gibson, S. and Hui, J. (2000) Modal parameter identification of a MEMS gyroscope, in Proceedings of the 2000 American Control Conference, Chicago, IL, pp. 1699–704.

19 Stone, H. and Kim, S. (2001) Microfluidics: basic issues, applications and challenges. *AIChE Journal*, **47** (6), 1250–4.

20 Shapiro, B. (2004) Workshop on control and system integration of micro- and nano-scale systems, in *Technical Report*, National Science Foundation, Washington, DC.

21 Shapiro, B. (2005) Control and system integration of micro- and nano-scale systems. *IEEE Control Systems Magazine*, **25** (2), 82–8.

22 Sitti, M. (2003) Workshop on future directions in nano-scale systems, dynamics and control, in *Technical Report*, National Science Foundation Workshop. Available on-line http://www.me.cmu.edu/faculty1/sitti/NSF_Workshop.html (accessed October 21, 2008).

23 Jie, D., Diao, X., Cheong, K.B. and Yong, L.K. (2000) Navier-Stokes simulations of gas flow in micro devices. *Journal of Micromechanics and Microengineering*, **10** (3), 372–9.

24 Karniadakis, G.E. and Beskok, A. (2002) *Micro Flows: Fundamentals and Simulation*, 2nd edn, Springer Verlag. New York, USA.

25 Jensen, K.F. (2001) Microreaction engineering–is small better? *Chemical Engineering. Science*, **56**, 293–303.

26 Stone, H.A., Stroock, A.D. and Ajdari, A. (2004) Engineering flows in small devices: microfluidics toward a lab-on-a-chip. *Annual Review of Fluid Mechanics*, **36**, 381–411.

27 Stroock, A.D., Dertinger, S.K.W., Ajdari, A., Mezic, I., Stone, H.A. and Whitesides, G.M. (2002) Chaotic mixer for microchannels. *Science*, **295**, 647–50.

28 Alfadhel, K. and Kothare, M.V. (2005) Microfluidic modeling and simulation of flow in membrane microreactors. *Chemical Engineering Science*, **60** (11), 2911–26.

29 Shen, C., Fan, J. and Xie, C. (2003) Statistical simulation of rarefied gas flows in microchannals. *Journal of Computational Physics*, **189**, 512–26.

30 Snyder, M.A., Vlachos, D.G. and Katsoulakis, M.A. (2003) Mesoscopic modeling of transport and reaction in microporous crystalline membranes. *Chemical Engineering Science*, **58**, 895–901.

31 Sirovich, L. (1987) Turbulence and the dynamics of coherent structures. *Quarterly of Applied Mathematics*, **XLV**, 561–71.

32 Finlayson, B.A. (1972) The method of weighted residuals and variational principles, with application in fluid mechanics, heat and mass transfer, in *Mathematics in Science and Engineering*, Vol. 87 (ed. R. Bellman), Academic Press, New York.

33 Comsol, A.B. and Natick, MA. (2001) FEMLAB Reference Manual, November.

34 Graham, M.D. and Kevrekidis, I.G. (1996) Alternative approaches to the Karhunen-Loeve decomposition for model reduction and data analysis. *Computers & Chemical Engineering*, **20**, 495–506.

35 Shvartsman, S.Y. and Kevrekidis, I.G. (1998) Nonlinear model reduction for control of distributed parameter systems: a computer assisted study. *AIChE Journal*, **44**, 1579.

36 Baker, J. and Christofides, P.D. (2000) Finitedimensional approximation and control of non-linear parabolic PDE systems. *International Journal of Control*, **73**, 439–56.

37 Ravigururajan, T.S. (1999) Two-phase flow characteristics of refrigerant flows in a microchannel heat exchanger. *Journal of Enhanced Heat Transfer*, **6** (6), 419–27.

38 Liang, Y.C. (2002) Proper orthogonal decomposition and its applications. *Journal of Sound and Vibration*, **252**, 527–44.

39 Loéve, M. (1955) *Probability Theory*, Van Nostrand, Princeton, NJ.

40 Cuta, J.M., McDonald, C.E. and Shekarriz, A. (1996) Forced convection heat transfer in parallel channel array microchannel heat exchanger, in Proceedings of the 1996 ASME International Mechanical Engineering Congress and Exposition, ASME, Atlanta, GA, November 17–22, vol. 338, pp. 17–23.

41 Stanley, R.S., Barron, R.F. and Ameel, T.A. (1997) Two-phase flow in microchannels, in Proceedings of the 1997 ASME International Mechanical Engineering Congress and Exposition, ASME, Dallas, TX, November 16–21, vol. 62, pp. 143–52.

42 Triplett, K.A., Ghiaasiaan, S.M., Abdel-Khalik, S.I., LeMouel, A. and McCord, B.N. (1999) Gas-liquid two-phase flow in microchannels. Part II: void fraction and pressure drop. *International Journal of Multiphase Flow*, **25** (3), 395–410.

43 Triplett, K.A., Ghiaasiaan, S.M., Abdel-Khalik, S.I. and Sadowski, D.L. (1999) Gas-liquid two-phase flow in microchannels. Part I: two-phase flow patterns. *International Journal of Multiphase Flow*, **25** (3), 377–94.

44 Banerjee, A., Arkun, T., Ogunnaike, B. and Pearson, R. (1997) Estimation of nonlinear systems using linear multiple models. *AIChE Journal*, **43**, 1204–26.

45 Kothare, M.V., Mettler, B., Morari, M., Bendotti, P. and Falinower, C.-M. (2000) Level control in the steam generator of a nuclear power plant. *IEEE Transactions on Control Systems Technology*, **8** (1), 55–69.

46 Murray-Smith, R. and Johansen, T.A. (eds) (1997) *Multiple Model Approaches to Modeling and Control*, Taylor and Francis.

47 Narendra, K.S. and Balakrishnan, J. (1997) Adaptive control using multiple models. *IEEE Transactions on Automatic Control*, **42** (2), 171–87.

48 Cass, S. (2001) MEMS in space. *IEEE Spectrum*, **38** (7), 56–61.

49 Park, S. and Horowitz, R. (2001) Adaptive control for Z-axis MEMS gyroscopes, in Proceedings of the 2001 American Control Conference, Arlington, VA, June, pp. 1223–8.

50 Krstic, M. and Smyshlyaev, A. (2007) Backstepping boundary control: a tutorial, in Proceedings of the 2007 American Control Conference, New York, NY, July 9–13, pp. 870–5.

51 Smyshlyaev, A. and Krstic, M. (2007) Adaptive boundary control for unstable parabolic PDEs–Part II: estimation-based designs. *Automatica*, **43** (9), 1543–56.

52 Smyshlyaev, A. and Krstic, M. (2007) Adaptive boundary control for unstable parabolic pdes–part iii: output feedback examples with swapping identifiers. *Automatica*, **43** (9), 1557–64.

53 Olmstead, W.E. (1980) Boundary controllability of the temperature in a long rod. *International Journal of Control*, **31** (3), 595–600.

54 Olmstead, W.E. and Schmitendorf, W.E. (1977) Optimal control of the end-temperature in a semi-infinite rod. *Journal of Applied Mathematics and Physics*, **28**, 697–706.

55 Olivei, A. (1974) Boundary modal control of the temperature of the melt for crystalgrowing in crucibles. *International Journal of Control*, **20** (1), 129–57.

56 Hanczyc, E.M. and Palazoglu, A. (1996) Use of symmetry groups in sliding model control of nonlinear distributed parameter systems. *International Journal of Control*, **63** (6), 1149–66.

57 Chakravarti, S. and Harmon Ray, W. (1999) Boundary identification and control of distributed parameter systems using singular functions. *Chemical Engineering Science*, **54**, 1181–204.

58 Lumley, J.L. (1970) *Stochastic Tools in Turbulence*, Academic Press, New York.

59 Rawlings, J.B. (2000) Tutorial overview of model predictive control. *IEEE Control Systems Magazine*, **20** (3), 38–52.

60 The MathWorks, Inc. (1992) Natick, Mass. MATLAB Reference Guide.

61 Bleris, L.G. and Kothare, M.V. (2005) Real-time implementation of model predictive control, in Proceedings of the 2005 American Control Conference, Portland, OR, June 8–10, pp. 4166–71.

62 Metrowerks Corporation (2004) CodeWarrior Development Studio, MPC5xx Edition, Austin, TX.

63 Karagianni, K., Chronopoulos, T., Tzes, A., Kousoulas, N. and Stouraitis, T. (1998) Efficient processor arrays for the implementation of the generalized predictive-control algorithm. *IEE Proceedings Control Theory and Applications*, **145** (1), 47–54.

64 Ling, K.V., Yue, S.P. and Maciejowski, J.M. (2006) An FPGA implementation of model predictive control, in Proceedings of the 2006 Americal Control Conference, Minneapolis, MN, June, pp. 1930–35.

65 Swartzlander, E.E. and Alexopoulos, A.G. (1975) The sign/logarithm number system.

IEEE Transactions on Computers, **24** (12), 1238–42.

66 Garcia, J.G., Arnold, M.G., Bleris, L.G. and Kothare, M.V. (2004) LNS architectures for embedded model predictive control processors, in Proceedings of the 2004 International Conference on Compilers, Architectures and Synthesis for Embedded Systems, Washington, DC, September, pp. 79–84.

Index

Microfabricated Power Generation Devices. Edited by Alexander Mitsos and Paul I. Barton
Copyright © 2009 WILEY-VCH Verlag GmbH & Co. KGaA, Weinheim
ISBN: 978-3-527-32081-3